Helena Khaliullina-Skultety

Regulation of Smoothened activity by Patched and lipoprotein lipids

Helena Khaliullina-Skultety

Regulation of Smoothened activity by Patched and lipoprotein lipids

Insights into the mechanism of inhibition

Südwestdeutscher Verlag für Hochschulschriften

Impressum/Imprint (nur für Deutschland/only for Germany)
Bibliografische Information der Deutschen Nationalbibliothek: Die Deutsche Nationalbibliothek verzeichnet diese Publikation in der Deutschen Nationalbibliografie; detaillierte bibliografische Daten sind im Internet über http://dnb.d-nb.de abrufbar.
Alle in diesem Buch genannten Marken und Produktnamen unterliegen warenzeichen-, marken- oder patentrechtlichem Schutz bzw. sind Warenzeichen oder eingetragene Warenzeichen der jeweiligen Inhaber. Die Wiedergabe von Marken, Produktnamen, Gebrauchsnamen, Handelsnamen, Warenbezeichnungen u.s.w. in diesem Werk berechtigt auch ohne besondere Kennzeichnung nicht zu der Annahme, dass solche Namen im Sinne der Warenzeichen- und Markenschutzgesetzgebung als frei zu betrachten wären und daher von jedermann benutzt werden dürften.

Verlag: Südwestdeutscher Verlag für Hochschulschriften GmbH & Co. KG
Dudweiler Landstr. 99, 66123 Saarbrücken, Deutschland
Telefon +49 681 37 20 271-1, Telefax +49 681 37 20 271-0
Email: info@svh-verlag.de

Approved by: Dresden, TU Dresden, Dissertation, 2010

Herstellung in Deutschland:
Schaltungsdienst Lange o.H.G., Berlin
Books on Demand GmbH, Norderstedt
Reha GmbH, Saarbrücken
Amazon Distribution GmbH, Leipzig
ISBN: 978-3-8381-2811-5

Imprint (only for USA, GB)
Bibliographic information published by the Deutsche Nationalbibliothek: The Deutsche Nationalbibliothek lists this publication in the Deutsche Nationalbibliografie; detailed bibliographic data are available in the Internet at http://dnb.d-nb.de.
Any brand names and product names mentioned in this book are subject to trademark, brand or patent protection and are trademarks or registered trademarks of their respective holders. The use of brand names, product names, common names, trade names, product descriptions etc. even without a particular marking in this works is in no way to be construed to mean that such names may be regarded as unrestricted in respect of trademark and brand protection legislation and could thus be used by anyone.

Publisher: Südwestdeutscher Verlag für Hochschulschriften GmbH & Co. KG
Dudweiler Landstr. 99, 66123 Saarbrücken, Germany
Phone +49 681 37 20 271-1, Fax +49 681 37 20 271-0
Email: info@svh-verlag.de

Printed in the U.S.A.
Printed in the U.K. by (see last page)
ISBN: 978-3-8381-2811-5

Copyright © 2011 by the author and Südwestdeutscher Verlag für Hochschulschriften GmbH & Co. KG and licensors
All rights reserved. Saarbrücken 2011

...dedicated to my parents, who always answered my questions...

Contrariwise, if it was so, it might be; and if it were so, it would be; but as it isn't, it ain't. That's logic.

Alice laughed. 'There's no use trying,' she said. 'One can't believe impossible things.'

'I daresay you haven't had much practice,' said the Queen. 'When I was your age, I always did it for half-an-hour a day. Why, sometimes I've believed as many as six impossible things before breakfast."

— Lewis Carroll

TABLE OF CONTENTS

1. Summary_____5
2. Abbreviations_____7
3. Introduction_____9

3.1 The morphogen Hedgehog in development, evolution and cancer

3.2 Routers of Hedgehog signaling

3.3 The paradigm of Smoothened signaling

3.4 Special features of Patched

3.5 Lipid links to Smoothened

3.6 Lipids and lipoproteins in Drosophila melanogaster

4. Scope of the thesis_____30
5. Results_____32

5.1 Lipophorin is required to reduce Smoothened accumulation on the basolateral membrane

5.2 Lipophorin particles directly affect Smoothened accumulation

5.3 The lipid contents of Lipophorin reduce basolateral Smoothened accumulation and reduce levels of Ci_{155}

5.4 The Sterol-Sensing Domain of Patched makes Lipophorin-derived lipids available for Smoothened repression

5.5 Patched recruits Lipophorin to early endosomes

5.6 The structural requirements of Patched

5.7 Patched diverts a subset of internalized Lipophorin to Patched-positive *endosomes*

5.8   *Mutation of the Patched Sterol-Sensing Domain perturbs lipid trafficking from Patched-positive endosomes*

5.9   *Lipids mobilized by Patched are derived from newly delivered Lipophorin*

5.10 *Mutation of the Patched Sterol-Sensing Domain perturbs Smoothened trafficking from Patched-positive endosomes*

5.11 *Purification of the active lipid species from Lipophorin-derived lipids*

6. Discussion _____ 68

6.1   *The ongoing mystery of Smoothened repression – clues so far*

6.2   *The Sterol-Sensing Domain of Patched regulates Smoothened trafficking*

6.3   *Endogenous Smoothened inhibitor is derived from Lipophorin particles by Patched*

6.4   *Role of Patched in lipid trafficking*

6.5   *Lipid candidates for Smoothened repression*

6.6   *Possibilities of Smoothened regulation*

6.7   *Further transduction of Smoothened signaling*

6.8   *Two possible modes of Smoothened signaling*

6.9   *Applications*

7.	**Supplements** _____	87
8.	**Material and methods** _____	93
9.	**Bibliography** _____	102
10.	**Acknowledgement** _____	116
11.	**Publication** _____	118

TABLE OF FIGURES

3. Introduction

Figure 3.1   Hedgehog is produced by the posterior compartment cells of the wing imaginal disc and signals by creating a concentration gradient in the tissue.

Figure 3.2   Schematic representation of the Hedgehog signaling pathway in Drosophila melanogaster.

Figure 3.3   Schematic representation of a bacterial RND permease, NPC-1 and Patched.

Figure 3.4   Schematic structure of Lipophorin particle.

Figure 3.5   Role of Lipophorin in Hedgehog transport.

5. Results

Figure 5.1   Lipophorin RNAi increases basolateral Smoothened accumulation.

Figure 5.2   Smoothened and Ci_{155} levels are not affected by dietary lipid depletion.

Figure 5.3   Lipophorin particles reverse Smoothened accumulation and Ci_{155} stabilization.

Figure 5.4   Lipophorin-derived lipids reduce basolateral Smoothened accumulation.

Figure 5.5   Lipophorin lipids reduce Ci_{155} levels.

Figure 5.6   Treatment with Lipophorin lipids does not affect Arrow levels.

Figure 5.7   Treatment with phosphatidylcholine or ergosterol does not affect basolateral Smoothened levels.

Figure 5.8   Lipophorin lipids induce translocation of Smoothened from the basolateral membrane to apical endosomes.

Figure 5.9   Lipophorin acts with Patched to influence Smoothened trafficking.

Figure 5.10   Patched induces Lipophorin accumulation in Rab5-positive endosomes.

Figure 5.11   The C-terminal region of Patched is essential to recruit Lipophorin.

Figure 5.12   Patched does not affect Lipophorin internalization, but decreases its degradation.

Figure 5.13   Patched diverts trafficking of a subset of internalized Lipophorin.

Figure 5.14   Mutation of the Patched Sterol-Sensing Domain prevents endosomal sterol efflux.
Figure 5.15   The specific effect of the Sterol-Sensing Domain of Patched on newly delivered sterol.
Figure 5.16   Mutation of the Patched Sterol-Sensing Domain traps Smoothened in Patched-positive endosomes.
Figure 5.17   Over-expression of Patched^{SSD} does not affect basolateral proteins Fasciclin III and Arrow.
Figure 5.18   Effects of the mutation in the Sterol-Sensing Domain of Patched on Smoothened trafficking.
Figure 5.19   Specific fraction of Lipophorin lipids reduces basolateral Smoothened accumulation.
Figure 5.20   Non-saponifiable Lipophorin-derived lipids reduce basolateral Smoothened accumulation.

6. Discussion

Figure 6.1   Schematized structure of Smoothened.
Figure 6.2   A model for Patched-mediated Smoothened destabilization.

7. Supplements

Figure 7.1   Protein-free derivation of Lipophorin lipids.
Figure 7.2   Major lipids present in Lipophorin.
Figure 7.3   Patched over-expression does not perturb endocytic compartments.
Figure 7.4   Expression levels of different Patched alleles.
Figure 7.5   LpR1 increases Lipophorin uptake, but does not affect Lipophorin degradation.
Figure 7.6   Patched decreases Lipophorn degradation and stabilizes it in early endosomes.
Figure 7.7   Mutation of the Sterol-Sensing Domain of Patched perturbs derivation of lipid cargo from internalized Lipophorin particles.
Figure 7.8   Localization of Smoothened in Patched-over-expressing cells.
Figure 7.9   Amount of Lipophorin-derived lipids necessary for reduction of basolateral Smoothened levels.

SUMMARY

SUMMARY

Hedgehog is a lipid-linked morphogen that is carried on lipoprotein particles and that regulates both patterning and proliferation in a wide variety of vertebrate and invertebrate tissues. Hyperactivity of Hedgehog signaling causes numerous forms of cancer. Hedgehog acts by binding to its receptor Patched, relieving the suppression of Smoothened and initiating Smoothened signaling. The mechanism by which Patched represses Smoothened has been unclear, but correlates with reduced Smoothened levels on the basolateral membrane. The structural homology of Patched with the Niemann-Pick-Type C1 protein and bacterial transmembrane transporters suggests that Patched might regulate lipid trafficking to repress Smoothened. However, no endogenous lipid regulators of Smoothened have yet been identified, nor has it ever been shown that Patched actually controls lipid trafficking.

This work shows that, in *Drosophila melanogaster*, the Sterol-Sensing Domain of Patched regulates Smoothened trafficking from Patched-positive endosomes. Furthermore, it demonstrates that Patched recruits internalized lipoproteins to Patched-positive endosomes. Thereby, Patched regulates the efflux of specific lipoprotein-derived lipids from this compartment via its Sterol-Sensing Domain and utilizes these lipids to destabilize Smoothened on the basolateral membrane.

We propose that Patched normally promotes Smoothened degradation and subsequently downregulates its activity by changing the lipid composition of endosomes through which Smoothened passes. For this purpose, Patched utilizes a specific lipid – possibly a modified sterol or sphingolipid – derived from lipoproteins. Further, we suggest that the presence of Hedgehog on lipoprotein particles inhibits utilization of their lipids by Patched.

2 ABBREVIATIONS

ABC	ATP-binding cassette
AMP	adenosine monophosphate
ATP	adenosine triphosphate
A/P	anteroposterior
ApoLI	Apolipophorin I
ApoLII	Apolipophorin II
Arr	Arrow
CBP	CREB-binding protein
Ci	Cubitus interruptus
CKI	Casein kinase I
Col	collier
Cos2	Costal 2
CREB	cyclic AMP-response element
DNA	desoxyribonucleic acid
Dpp	Decapentaplegic
En	engrailed
FasIII	Fasciclin III
Fu	Fused
GFP	green fluorescent protein
Gli	Glioma-associated oncogene homologue
GSK3β	Glycogen synthase kinase-3 β
HDL	high-density lipoprotein
Hh	Hedgehog
IDL	intermediate-density lipoprotein
Iro	iroquois
LDL	low-density lipoprotein
Lpp	Lipophorin
LpR	LDL receptor homologue
MDR	multidrug resistance
NPC-1	protein encoded by the Niemann Pick Type C 1 disease gene
PC	phosphatidylcholine

PE		phosphatidylethanolamine
PKA		protein kinase A
PI		phosphatidylinositol
PS		phosphatidylserine
PCE		phosphatidylcholine-ergosterol
Ptc		Patched
RNA		ribonucleic acid
RNAi		RNA interference
RND		Resistance Nodulation Division
SCAP		SREBP cleavage-activating protein
Smo		Smoothened
SREBP		sterol regulatory element-binding protein
SSD		Sterol-Sensing Domain
Sufu		Suppressor of Fused
TAG		triacylglycerol
TLC		thin layer lipid chromatography
UAS		upstream activator sequence
VLDL		very low-density lipoprotein
wt		wild type

INTRODUCTION

3 INTRODUCTION

As an organism develops through life, it faces a crucial challenge – it has to establish an enormous palette of various tissues, each of them unique in its function and regulation. To achieve this, both vertebrate and invertebrate developmental processes need to be strictly controlled spatially and temporally on the cellular level. Thereby, morphogens are important performers of this pivotal role. They are regulatory molecules, which govern the pattern of tissue development and, in particular, the positioning of the specialized cell types within a tissue. Generally speaking, morphogens elicit numerous and diverse cellular responses, dependent on their specific concentration that reaches the cell. Normally, a morphogen is produced at a discrete source and gradually spreads throughout the tissue to create a concentration gradient. As a result, different cell fates are defined along such a gradient, each corresponding to a distinct morphogen concentration (Teleman et al., 2001; Martinez Arias, 2003).

3.1 The morphogen Hedgehog in development, evolution and cancer

Hedgehog is one of the most important morphogens regulating growth and differentiation and it is highly conserved throughout the majority of the animal kingdom. Hedgehog was first discovered in *Drosophila melanogaster*, where it functions to regulate the body segmentation in embryos. There, loss of Hedgehog signaling causes *Drosophila* embryos to develop abnormal cuticular spikes, inspiring the name "hedgehog" (Machold et al., 2003; Ingham and Placzek, 2006). Generally, Hedgehog controls the development of most organs in both vertebrates and invertebrates (McMahon et al., 2003; Ingham and Placzek, 2006). For instance, in the fly wing precursor – the wing imaginal disc – Hedgehog ensures the correct expression of the target genes and subsequently proper development of the wing (Tabata and Kornberg, 1994; Strigini and Cohen, 1997). In the third instar larva, the wing disc is an epithelial sheet and develops as a sack-like structure that will give rise to the adult wing (**Fig. 3.1 A**). The Hedgehog protein is secreted by the cells of the posterior compartment and is transported to the adjacent anterior compartment, where it initiates an instructive signal to produce discrete cellular states in a concentration-dependent manner (**Fig. 3.1 B**) (Strigini and Cohen, 1999; Tabata, 2001; Teleman et al., 2001).

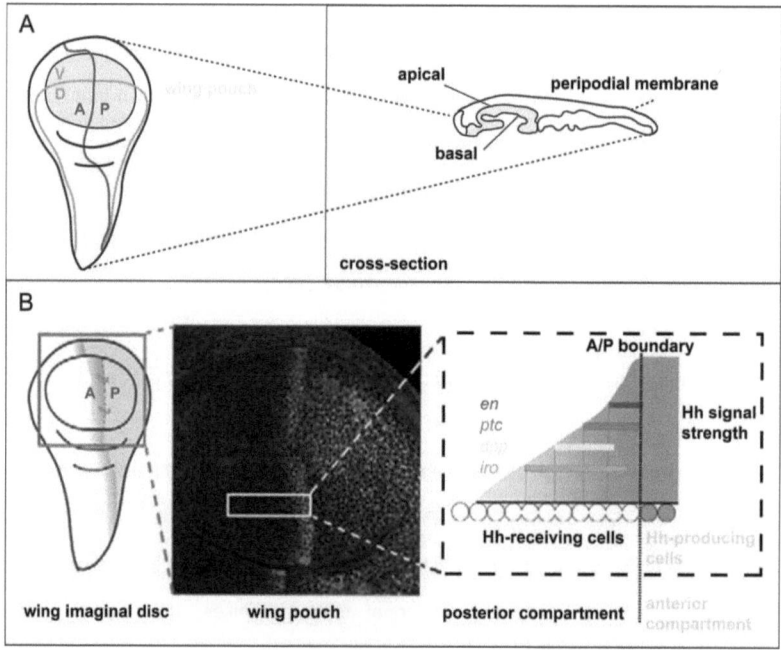

Figure 3.1 **Hedgehog is produced by the posterior compartment cells of the wing imaginal disc and signals by creating a concentration gradient in the tissue.**
(A) Schematic structure (left panel) and a cross-section (right panel) of a wing imaginal disc of a third instar larva. The wing disc is an epithelial sheet of cells covered by the peripodial membrane at the apical side. The wing pouch (yellow) gives rise to the adult wing blade and is patterned into

a dorsal (D), ventral (V), anterior (A) and posterior (P) compartment (adapted from (Butler et al., 2003). (B) A schematic of *Drosophila* wing imaginal disc staining (adapted from (Callejo et al., 2006). Hedgehog (Hh) is expressed in the posterior compartment cells of the wing imaginal disc and is marked by green staining. Patched (Ptc) expression is marked by red staining that appears as a stripe adjacent to the A/P boundary. Hh is spread to the receiving cells of the anterior compartment, where it creates a concentration gradient. There, distinct concentrations induce transcription of specific target genes, indicated by differently coloured bars – *engrailed* (*en*, blue) and *ptc* (*ptc*, pink) are high threshold targets, *decapentaplegic* (*dpp*, yellow) is a medium threshold target, and *iroquois* (*iro*, orange) is a low threshold target (adapted from (Wendler et al., 2006).

In vertebrates, genome duplication has given rise to multiple Hedgehog genes, among which Sonic Hedgehog is the most well studied (Echelard et al., 1993; Krauss et al., 1993; Riddle et al., 1993; Chang et al., 1994; Roelink et al., 1994). Nevertheless, the mechanism and the components of the signal transduction by the fly and mammalian Hedgehog proteins have remained largely conserved throughout evolution, exhibiting one of the major developmental regulator processes in the whole animal kingdom (Goodrich et al., 1996; Murone et al., 1999a).

Playing a crucial role in the embryonic patterning and growth, Hedgehog signaling pathway remains likewise important for the adult organism. There, it participates in the hematopoiesis and maintenance of certain stem cell populations (Baron, 2003). Thus, Sonic Hedgehog signaling regulates the proliferation and persistence of the stem cell niches in the adult mammalian brain and several endoderm-derived epithelia, for example, the stomach, intestine, and pancreas (Machold et al., 2003; Palma et al., 2005; van den Brink, 2007; Ishizuya-Oka and Hasebe, 2008). Further, in vertebrate skin, Sonic and Desert Hedgehog signal to ensure epidermal stem cell maintenance (Zhou et al., 2006).

Regarding these potent biological activities, it is not surprising that misregulation of the Hedgehog signaling in adult life can be associated with various malformations.

Thus, excessive Hedgehog pathway activity – caused by over-production of Hedgehog ligand by tumor cells or by mutations of signaling pathway components – provokes development of a large variety of tumors, such as medulloblastoma, basal-cell carcinoma, prostate cancer, pancreatic adenocarcinoma, tumors of the digestive tract and certain small-cell lung cancers (Ruiz i Altaba et al., 2002; Pasca di Magliano and Hebrok, 2003; Dellovade et al., 2006). Moreover, continued Hedgehog activity is often required for maintenance, in addition to initiation, of tumor growth – Hedgehog regulates cancer stem cell self-renewal and proliferation indirectly by influencing the surrounding stroma, which provides a more favorable environment for tumor expansion (Clement et al., 2007; Yauch et al., 2008). Thus, regarding these fatal consequences caused by inadequate regulation, Hedgehog signal propagation needs to be strictly controlled to ensure an appropriate feedback. Notably, Hedgehog signaling pathway comprises a large potential for regulation that can be achieved at multiple levels – the network of its signal transductors is outlined in the next section.

3.2   Routers of Hedgehog signaling

Remarkably, Hedgehog can trigger a large diversity of signal outcomes, employing the same signaling pathway. For instance, in *Drosophila* wing imaginal disc, different Hedgehog concentrations and/or varying signal duration provide a basis for differential sensitivity of the responding cells of the anterior compartment. This results in a patterned expression of the target genes – highest concentrations of Hedgehog induce the transcription of *collier* and *patched*, intermediate Hedgehog concentrations promote transcription of *decapentaplegic* and low levels of *patched* expression and low amounts of Hedgehog lead to expression of *iroquois*, at the same time repressing the transcription of *collier*, *patched* and *decapentaplegic* (**Fig. 3.1**). Accordingly, activation or repression of these genes specifies the cellular identity throughout the developmental process. This section highlights the router points in Hedgehog signal transduction, which serve as cellular rheostats to translate variable Hedgehog concentrations to different levels of pathway activation and target gene transcription (see also **Fig. 3.2**).

Once Hedgehog reaches the responding cell, it associates with its receptor Patched, a multipass transmembrane protein, which is only expressed by the cells of the anterior compartment (Ingham et al., 1991; Tabata and Kornberg, 1994; Chen and Struhl, 1996; Marigo et al., 1996; Stone et al., 1996). In addition, in Drosophila, two other transmembrane proteins – Interference Hedgehog and Brother of Ihog – have recently been shown to play a role of co-receptors in this process, interacting with Patched to facilitate Hedgehog binding (Williams et al., 2008). Consequently, Hedgehog binding to Patched relieves the inhibitory effect of Patched on Smoothened, a 7-transmembrane domain protein (Alcedo et al., 1996; Marigo et al., 1996; van den Heuvel and Ingham, 1996a; Alcedo and Noll, 1997; Chen and Struhl, 1998; Murone et al., 1999b). Once Smoothened is derepressed, it activates the signaling pathway by modulating the levels and activity of the transcription factor Cubitus interruptus (Ci) (Alexandre et al., 1996; Von Ohlen and Hooper, 1997). Ci is a member of the Glioma-associated oncogene homologue (Gli) transcription-factor family and has three mammalian homologues.

Modulation of the activation state of Ci serves as the main controlling checkpoint for the specific response to Hedgehog signal. Ci is a bifunctional transcription factor with variable modes of activity – it contains both repressor and activator domains that flank a central DNA-binding zinc-finger domain. In the absence of Smoothened signaling, the full-length Ci_{155} undergoes proteolytic cleavage, primed by its phosphorylation by three kinases – Protein kinase A (PKA), Glycogen synthase kinase-3 (GSK3β) and Casein kinase I (CKI) and mediated by the ubiquitin ligase pathway. A physical basis for these interactions is provided by the association of Ci_{155} with the scaffolding protein Costal 2, which recruits these kinases to mediate Ci_{155} processing (Lum et al., 2003; Zhang et al.,

2006). This yields a truncated form of the protein – Ci_{75}, which acts exclusively as a repressor for Hedgehog target genes. Smoothened activity prevents Ci degradation and stabilizes the full-length form of the protein – Ci_{155} – in the cytoplasm. Thus, depletion of the repressor form Ci_{75} relieves inhibition of some Hedgehog target genes, generating one of the possible signaling responses.

Further, stabilized Ci_{155} can be activated, possibly by a post-translational modification, translocates to the nucleus and enhances the transcription of Hedgehog target genes, facilitating the high-level response and enabling other possible signaling scenarios (Aza-Blanc and Kornberg, 1999; Lefers et al., 2001; Nybakken and Perrimon, 2002a; Lum and Beachy, 2004; Kalderon, 2005; Smelkinson et al., 2007).

Positive regulation of Hedgehog signaling is supported by Fused, a serine-threonine kinase, which interacts with Costal 2 and prevents Ci_{155} degradation (Jia et al., 2003; Ruel et al., 2003). On the other hand, the action of Fused is opposed by the effector protein Suppressor of Fused (Sufu), which associates with Ci_{155}, thereby retaining Ci_{155} in the cytoplasm and preventing its nuclear translocation and target gene activation (Methot and Basler, 2000; Merchant et al., 2004; Kalderon, 2005). In *Drosophila*, Hedgehog signaling can be additionally attenuated by the protein Roadkill, which targets Ci_{155} for Cullin3-mediated ubiquitinylation and degradation (Kent et al., 2006).

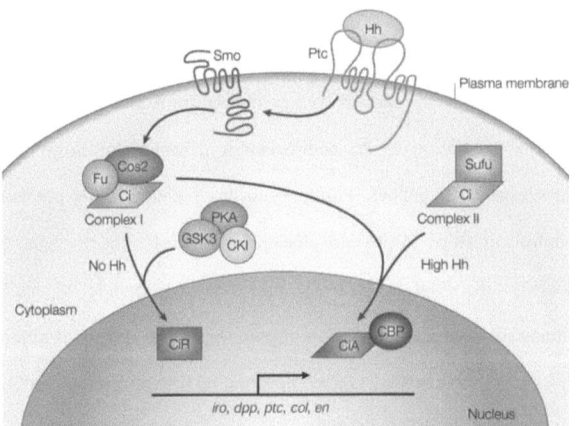

Figure 3.2 **Schematic representation of the Hedgehog signaling pathway in *Drosophila melanogaster*.**
The transcription factor Cubitus interruptus (Ci) is held in the cytoplasm by Complex I, which consists of the serine/threonine kinase Fused (Fu) and the kinesin-related protein Costal 2 (Cos2), or by Complex II, which includes Suppressor of Fused (Sufu). In the absence of the Hh ligand, Ptc inhibits Smoothened (Smo). Complex I promotes the phosphorylation of Ci by Protein kinase A (PKA), Glycogen synthase kinase 3 β (GSK3β) and Casein kinase I (CKI), and promotes subsequent processing of Ci to the transcriptional repressor, CiR. Low concentrations of Hh function through Ptc and then Smo to block the production of CiR. Higher concentrations of Hh function through Ptc, Smo and Complexes I and II to release an active form of Ci – CiA – that stimulates transcription through the co-activator CBP. Transcriptional targets include *dpp*, *ptc*, *en*, *collier (col)*, *iro* and many other genes (adapted from (Hooper and Scott, 2005).

Finally, the activity of Hedgehog signaling pathway can also be modulated at the level of Smoothened. It has been postulated recently that, upon derepression, *Drosophila* Smoothened becomes hyperphosphorylated by several kinases (Jia et al., 2004; Zhang et al., 2004; Apionishev et al., 2005). In this way, increasing phosphorylation gradually increases Smoothened activity, translating graded Hedgehog signals into distinct responses (Zhao et al., 2007). Although the phosphorylation residues of Smoothened are not conserved in vertebrates, it has recently been shown that mammalian Smoothened is influenced by the activity of G-protein coupled receptor kinases and might therefore undergo similar modifications (Philipp et al., 2008). Thus, a signal provided by Hedgehog can be progressively translated into a step-by-step activation mechanism, bearing in its complexity a potential for large variability of commands defining cellular identities. Furthermore, the outlined router points can serve as important controlling steps to provide efficient negative feedback of the Hedgehog signaling, especially important considering the extensive palette of malformations caused by its inadequate regulation. In this regard, however, the most straightforward control of Hedgehog pathway activity is accomplished by the inhibitory action of Patched on Smoothened. The nature of this inhibition has remained the most ambiguous – and probably the most intriguing – mystery of the Hedgehog signaling.

3.3 The paradigm of Smoothened activity

Smoothened belongs to the superfamily of seven-transmembrane receptors; its N-terminal cystein-rich extracellular domain and the membrane-spanning regions are highly conserved across phyla (Quirk et al., 1997; Pitcher et al., 1998).

Initially, the mechanism of Smoothened inhibition by Patched in the absence of Hedgehog signal was thought to be of physical nature. Indeed, when vastly overexpressed, Patched was shown to weakly interact with the N-terminal extracellular domain of Smoothened (Stone et al., 1996). However, this interference appears to be of an artificial nature, since it could neither be detected under physiological conditions, nor were Patched and Smoothened observed to colocalize in resting cells (Johnson et al., 2000; Taipale et al., 2002; Zhu et al., 2003; Torroja et al., 2004). Instead of that, it was demonstrated that Patched represses Smoothened catalytically, because the inhibition occurs at a stochiometry of up to 1:50 (Taipale et al., 2002).

Furthermore, the activity status of Smoothened seems to correlate remarkably with its subcellular localization. In general, trafficking of the pathway components and the subcellular organization of Hedgehog signaling have repeatedly been observed to be determining. Thus, in vertebrate cells the signaling components localize to a special membrane compartment – the primary cilium (Corbit et al., 2005; Haycraft et al., 2005; Huangfu and Anderson, 2005; Huangfu and Anderson, 2006; Rohatgi et al., 2007). Primary cilia are plasma membrane projections, capable of detection and transduction of various extracellular signals (Gerdes et al., 2009). Movement of the pathway components into and out of the cilia regulates the activity of signaling. Accordingly, the binding of Hedgehog to Patched directs Patched out of the primary cilium and facilitates the translocation of Smoothened into this compartment (Corbit et al., 2005; Rohatgi et al., 2007). Also, the Gli transcription factors localize to the cilium upon pathway activation (Haycraft et al., 2005). In the absence of Hedgehog, Smoothened is detected in endocytic compartments and at the plasma membrane. Both these subcellular pools are thought to contribute to the ciliary movement of Smoothened upon the induction of signaling (Milenkovic et al., 2009).

Drosophila cells do not possess primary cilia, therefore extrapolations from studies in vertebrates have been ambiguous. Nevertheless, modules of the *Drosophila* Hedgehog pathway show similar affiliation to specific subcellular complexes. Thus, as already described earlier, the scaffolding protein Costal 2 binds the pathway components Smoothened, Fused, Ci and the kinases PKA, CKI and GSK3β to ensure their direct interactions.

Furthermore, in analogy to the mammalian system, a similar switch between subcellular compartments in response to the signal is observed in *Drosophila* wing imaginal disc cells. Here, the repressive action of Patched in the absence of Hedgehog results in decreased Smoothened stability and inhibits its accumulation on the basolateral membrane; vice versa, stimulation of signaling by Hedgehog targets Smoothened to the basolateral membrane (Denef et al., 2000; Nakano et al., 2004). Interestingly, Smoothened localization and its activation are tightly linked in a codependent way – thus, overexpression of Smoothened to non-physiological concentrations leads to accumulation of Smoothened at the cell surface, which alone suffices to activate the pathway (Hooper, 2003; Zhu et al., 2003; Zhang et al., 2004).

Another factor contributing to the regulation of Smoothened activity is its phosphorylation, as already outlined above. The cytoplasmic tail of *Drosophila* Smoothened contains a cluster of Ser/Thr residues, phosphorylation of which is both necessary and sufficient to activate Smoothened and induce signaling (Jia et al., 2004; Zhang et al., 2004; Apionishev et al., 2005). The mechanism for this induction has been suggested to correlate with a significant change in Smoothened conformation. Thus, in the absence of phosphorylation, multiple arginine clusters in the cytoplasmic Smoothened tail are thought to keep the protein in an inactive conformation via electrostatic interactions. Phosphorylation of the consensus sites gradually antagonizes this inhibition and facilitates progressive increase in Smoothened accumulation and activity (Zhao et al., 2007). This observation is not surprising, since it correlates well with the common properties of the seven-transmembrane receptors. Thus, most of them undergo changes in conformation when activated, which is achieved via phosphorylation through the recruited effectors. Interestingly, these events are also followed by a change in subcellular localization, amounts and/or signaling properties of the receptor (Pitcher et al., 1998). These findings provide additional support for the hypothesis, that Smoothened activity may primarily be regulated through its subcellular trafficking.

A large number of interesting findings further supports the development of this idea. For instance, it has been discovered that vertebrate Smoothened signaling can be artificially repressed or activated by a large variety of small lipophilic compounds such as the plant sterol derivative cyclopamine. Cyclopamine and its relative jervine are naturally occurring steroidal alkaloids of the corn lily *Veratrum californicum*, which, when fed to sheep, cause developmental malformations such as holoprosencephaly and cyclopia, due to defects in Hedgehog signaling (Keeler, 1978; Keeler and Balls, 1978). These and other small hydrophobic molecules have been shown to inhibit Smoothened through direct binding to its transmembrane helices and induce Hedgehog loss-of-function phenotypes (Frank-Kamenetsky et al., 2002; Chen et al., 2002a; Chen et al., 2002b). On the other hand, diverse hydrophobic compounds, natural – such as oxysterols – as well as artificial, have been demonstrated to act as agonists on Hedgehog signaling. However, whereas some of them apparently bind to the same binding site of Smoothened as cyclopamine, others do not appear to interact with Smoothened directly (Frank-Kamenetsky et al., 2002; Corcoran and Scott, 2006; Dwyer et al., 2007).

Furthermore, these compounds have been observed to influence the trafficking of vertebrate Smoothened, whereas the inhibition of its activity seemed to correlate with its depletion form the primary cilia. However, although ciliary localization of Smoothened is required for the activation of signaling, it is not sufficient, since treatment with the antagonist cyclopamine causes Smoothened translocation into cilia but inhibits its signaling activity. Other Smoothened antagonists appear to act by blocking ciliary localization, suggesting that Smoothened activity might be regulated at multiple steps by different ligands (Aanstad et al., 2009; Rohatgi et al., 2009; Wang et al., 2009).

In summary, these findings support the hypothesis that Smoothened activity can be regulated through its subcellular trafficking, which, in turn, could be influenced by certain endogenous lipophilic compounds, possibly related to the artificial antagonists mentioned above.

3.4 Special features of Patched

Whereas the hypothesis proposing Smoothened being regulated by small molecule compounds was favored by the studies described above, the role of Patched in this process has not been well understood. A clue came from the study of conserved motifs in Patched (Fig. **3.3**), which demonstrated that this protein is homologous to the prokaryotic permeases of the resistance-nodulation division (RND) family (Tseng et al., 1999; Ioannou, 2001; Taipale et al., 2002). The members of this family are efflux-pump proteins associated with multidrug-resistance. They are organized as tripartite systems, which facilitate the transport of various substrates across the membrane bilayer – a process driven by the proton motive force (Eswaran et al., 2004).

Another interesting structural feature of Patched is its Sterol-Sensing Domain (SSD), found in proteins, which can directly bind cholesterol, such as SCAP (SREBP cleavage-activating protein) and NPC-1 – the protein encoded by the Niemann Pick Type C 1 disease gene (Kuwabara and Labouesse, 2002; Ohgami et al., 2004; Radhakrishnan et al., 2004).

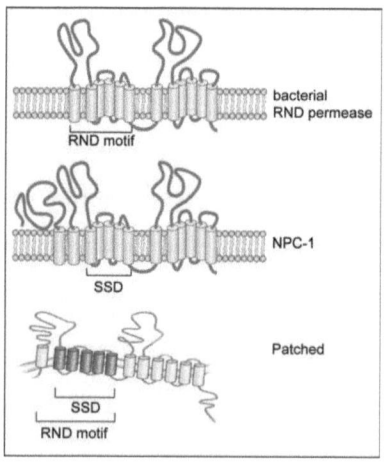

Figure 3.3 **Schematic representation of a bacterial RND permease, NPC-1 and Patched.**
The RND motif of the permease is known as the RND signature and is repeated twice. Both, Ptc and NPC-1 contain the RND signature and, furthermore, share the Sterol-Sensind Domain (SSD) (adapted from (Ioannou, 2001).

Both these motifs are of crucial importance for the repressive function of Patched, since mutations in either of them results in disability of Patched to inhibit Smoothened. More than that, the activity of the SSD of Patched seems to be implicated in vesicular trafficking, residing the Patched^{SSD} protein in the endosomes and suggesting a possible mechanistic link to the regulation of Smoothened activity (Martin et al., 2001; Strutt et al., 2001; Johnson et al., 2002). However, how exactly the SSD function is involved in Smoothened repression has not been demonstrated clearly yet. Interestingly, the function of NPC-1, the nearest relative of Patched proteins, has been established recently. NPC-1 promotes the mobilization of glycosphingolipids and LDL-derived cholesterol from late endosomes to other cellular membranes in both *Drosophila* and vertebrates and has been postulated to act as a transmembrane molecular pump (Ikonen and Holtta-Vuori, 2004; Mukherjee and Maxfield, 2004).

These correlations led to a suggestion that Patched, via its RND and SSD motifs, might similarly facilitate the transport of a small hydrophobic compound, which would influence Smoothened trafficking to repress its signaling activity. However, it has neither been shown that Patched can actually regulate lipid trafficking nor has an endogenous lipid been found to regulate Smoothened activity *in vivo* yet.

3.5   Lipid links to Smoothened

The central mystery of the Hedgehog pathway so far has been the regulatory step between Patched and Smoothened. As already mentioned above, the most favored theory suggests that Patched acts through a lipophilic modulator to repress Smoothened. This can be achieved through binding of such an inhibitor to Smoothened, inducing a conformational change that would promote deactivation of Smoothened – a possibility supported by a considerable susceptibility of Smoothened to regulation by various small molecule synthetic compounds. Alternatively, the inhibitory lipid could affect Smoothened activity by altering its subcellular trafficking and promoting its degradation. This does not necessarily imply a direct binding of the inhibitory compound to Smoothened and is especially supported by the fact that *Drosophila* Smoothened cannot be affected by binding of cyclopamine or some other known vertebrate Smoothened antagonist (Taipale et al., 2000). One way or the other, these hypotheses led to a large number of ongoing explorations focused to reveal an inhibitory lipid *in vivo*.

Thus, in consideration of the first possibility, one study suggested vitamin D3 being a direct Smoothened inhibitor based on its natural occurrence and its ability to compete with cyclopamine for the binding to mammalian Smoothened (Bijlsma et al., 2006). However, physiological role for vitamin D3 as an endogenous Hedgehog pathway inhibitor is questioned by many facts. First of all, its biosynthesis requires ultraviolet light (DeLuca, 2004), which is notably absent in the environment of developing mammalian embryos. In addition, vitamin D3 levels show strong variations between individuals and during lifetime, which would lead to major instabilities in Hedgehog signaling (Bischof et al., 2006). Vitamin D3-deficient individuals do not show increased cancer susceptibility, as would be expected if Smoothened were overactive (Xie and Bikle, 1998). In fact, vitamin D3 may even have an antiproliferative and proapoptotic effect on cancer cells (Schwartz and Skinner, 2007). Moreover, cattle that feed on vitamin D3-laden South American egg plants, *Solanum malacoxylon*, develop pathologic calcification (Wasserman et al., 1976) but do not show any developmental malformations that would resemble the Hedgehog-related defects seen so strikingly in *Veratrum*-poisoned sheep, in which Smoothened is overinhibited. In addition, the mentioned study also describes an effect of 7-dehydrocholesterol, the precursor to pro-vitamin D3, thus questioning the specific requirement for vitamin D3.

Furthermore, regulation of Smoothened activity through its trafficking has been supported by another study in vertebrate cells. There, Patched was speculated to affect Smoothened trafficking – and thereby its activity – by modifying the membranes of cellular compartments where Smoothened is present (Incardona et al., 2002). In general, protein trafficking has repeatedly been brought into correlation with the membrane lipid composition of the subcellular compartments. Thus, accumulation of certain lipid species, such as sphingolipids and cholesterol, contributes to formation of specific membrane domains – lipid rafts, which are thought to alter size and composition of membrane-associated protein sets, favoring recruitment of specific effectors. Thus, lipid rafts actively participate in intracellular membrane trafficking and protein sorting (Simons and Ikonen, 1997; Schuck and Simons, 2004). Interestingly, NPC-1 has been brought in correlation with rafts and has been shown to maintain raft lipid distribution employing its SSD (Chavrier et al., 1991; Simons and Gruenberg, 2000). Furthermore, lipid accumulation in endosomes of NPC-1 mutant cells has been observed to influence the activity of Rab7 (Lebrand et al., 2002), Rab9 (Ganley and Pfeffer, 2006) and Rab4 (Choudhury et al., 2004), perturbing degradation and recycling. These observations present yet another link between the cellular trafficking routes and the lipid environment of the subcellular compartments, providing interesting hints as to how Smoothened activity might be regulated by Patched. However, it has not been investigated in detail yet whether Patched is indeed able to regulate subcellular lipid trafficking, and this possibility needs to be established further.

3.6   Lipids and lipoproteins in Drosophila melanogaster

Vertebrate and invertebrate systems share many similarities regarding the principal lipid functions and their structural characteristics. However, *Drosophila* cells and tissues show some fundamental specificity in their lipid composition. First, it is worth to mention that *Drosophila*, like many insects, is a sterol auxotroph and relies on the dietary source to maintain the requirements for sterols. Second, *Drosophila* cells differ from the majority of vertebrate cells in their phospholipid as well as sphingolipid composition (Jones et al., 1992; Wiegandt, 1992; Rietveld et al., 1999; van Meer et al., 2008). Furthermore, whereas mammalian cells use triacylglycerol as their main storage lipid, *Drosophila* cells utilize diacylglycerol for this purpose (Arrese et al., 2001).

Once taken up from the dietary source, nutritional lipids need to be delivered to the various tissues in order to be utilized. Lipids are hydrophobic and hardly soluble in aqueous environment, such as circulation system, extracellular spaces and the subcellular compartments. Thus, to facilitate their transport, lipids associate with specific carrier proteins in the plasma to form lipoprotein particles, which can be easily carried to the receiving tissues. This is a widespread mechanism in both vertebrate and invertebrate systems and employs apolipoproteins in particular to form lipoprotein particles.

Lipoprotein particles are constituted of a core of triacylglycerols, sterols and sterol esters, which is surrounded by a phospholipid monolayer and scaffolded by the apolipoprotein (**Fig. 3.4**). Mammals possess five different types of lipoproteins, which have been classified based on their density as chylomicrons, very low-density lipoprotein (VLDL), intermediate-density lipoprotein (IDL), low-density lipoprotein (LDL) and high-density lipoprotein (HDL) (Shelness and Sellers, 2001; Vance and Vance, 2002).

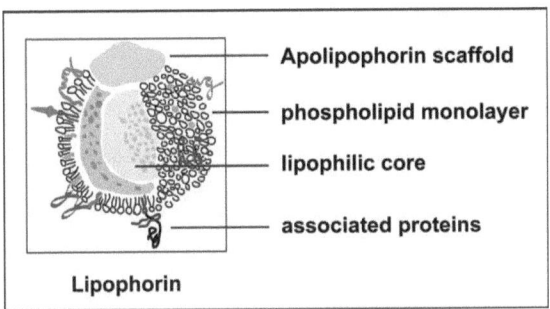

Figure 3.4 **Schematic structure of Lipophorin particle.**
Apolipophorin protein moiety scaffolds a lipophilic core mainly containing triacylglycerols, sterols and sterol esters, surrounded by a phospholipids monolayer. Various proteins, e.g. Hh, associate with Lipophorin (Lpp) via their lipophilic anchors.

In insects, lipids are also transported in lipoproteins. In several aspects, however, the insect system is very different from that in mammals (Beenakkers et al., 1985; Shapiro et al., 1988; Ryan, 1990a). Thus, in insects, lipid trafficking implicates one multifunctional transport vehicle, the major lipoprotein particle Lipophorin (Van der Horst, 1990; Sundermeyer et al., 1996). The Lipophorin particle has a molecular mass in the range of 500-600 kDa and is typically comprised of two glycosylated integral protein consituents, apolipophorin I and II, which scaffold a core of mixed lipids, surrounded by a phospholipid monolayer. Apart from sterols, sterol esters and hydrocarbons, the major lipid component carried by Lipophorin is diacylglycerol, in contrast to triacylglycerol in the mammalian systems (Ryan et al., 1990b).

Lipophorin has been characterized as a lipid shuttle, which is able to selectively load and unload lipid cargo at different target tissues (Van der Horst, 1990; Stanley and Nelson, 1993). In *Drosophila*, exclusively the cells of the fat body – an organ analogous to mammalian liver and adipose tissue, synthesize Apolipophorin. It is made as a proapolipoprotein precursor, which is posttranslationally processed and assembled into a Lipophorin particle (Kutty et al., 1996; Smolenaars et al., 2005). Subsequently, Lipophorin enters the systemic circulation and is transported via the hemolymph to the target tissues, where it is internalized. For instance, the cells of the wing imaginal disc of *Drosophila* express five different lipoprotein receptors – Megalin, Lrp1, Arrow and the two LDL receptor homologues LpR1 and LpR2 – which are responsible for the uptake of Lipophorin in the tissue (Khaliullina et al., 2009). Finally, internalized Lipophorin enters the endocytic route and is sequestered, making its lipid cargo available for further cellular needs (Tufail and Takeda, 2009).

The density of Lipophorin constitutes 1.12 g/ml and resembles that of HDL (Pho et al., 1996). This fact indicates a high protein/lipid ratio, which correlates with another important function of Lipophorin apart from its role in nutrition. Thus, Lipophorin has been shown to transport various proteins such as glypicans, which associate with Lipophorin via their hydrophobic anchors and/or heparan-sulfate moieties (Nybakken and Perrimon, 2002b; Eugster et al., 2007).

The ability of Lipophorin to serve as a protein carrier platform provides an interesting link with the principal feature of Hedgehog family proteins, which is their lipid modification (Mann and Beachy, 2004; Peters et al., 2004). The mature Hedgehog protein is synthesized as a precursor that undergoes a series of posttranslational modifications, leading to covalent attachment of a cholesterol moiety at its C-terminus and a palmitic acid at its N-terminus (Porter et al., 1996a; Pepinsky et al., 1998). Although the hydrophobic nature of Hedgehog modifications confers high affinity for cell membranes, this morphogen can only be efficiently secreted and signal normally in the presence of its lipid modifications (Porter et al., 1996b; Peters et al., 2004; Eaton, 2006; Guerrero and Chiang, 2007). This results from the association of Hedgehog with the extracellular matrix proteoglycans and, during its long-range movement, with the lipoproteins – Lipophorin in *Drosophila* (Panáková et al., 2005; Callejo et al., 2006; Eugster et al., 2007; Neumann et al., 2007).

In this way, Lipophorin also serves as a signaling platform, ensuring the delivery of messenger proteins to their target tissue. Thus, in *Drosophila* wing imaginal disc, Hedgehog secreted by cells of the posterior compartment is carried on Lipophorin particles to the responding cells of the anterior compartment (**Fig. 3.5** A). Fat body-driven knock-down of Lipophorin levels in the whole organism results in narrowed expression of long-range Hedgehog target genes (**Fig. 3.5** B and (Panáková et al., 2005).

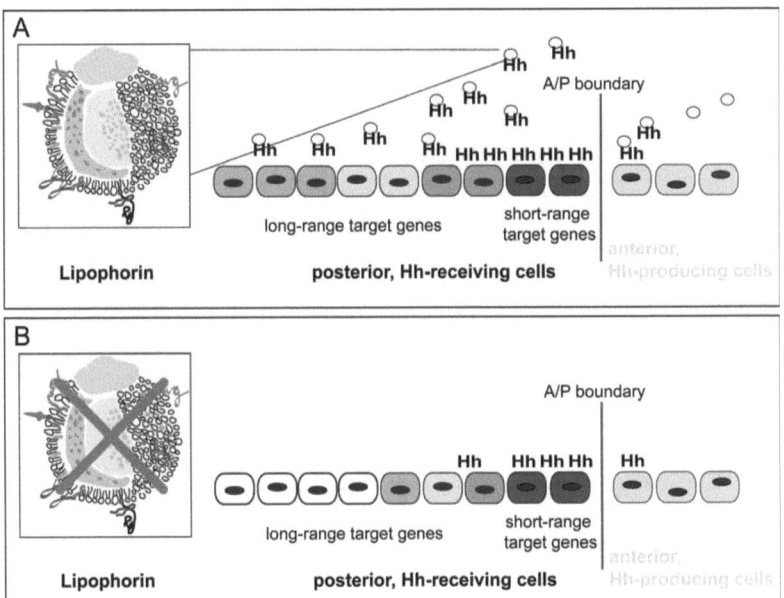

Figure 3.5 **Role of Lipophorin in Hedgehog transport.**
(A) Hh is produced in the posterior compartment cells (indicated in green) and associates with Lpp in order to be transported over long-range distances in the anterior compartment containing

Hh-receiving cells. There, Hh induces transcription of short- and long-range target genes, depending on its concentration (indicated by blue, pink, yellow and orange). (B) Absence of Lpp interferes with long-range Hh transport, resulting in a narrowed expression of long-range target genes (Panáková et al., 2005).

In summary, Lipophorin presents itself as an important regulator of Hedgehog function, since its role as a transport vehicle ensures the correct formation of the morphogen gradient in the responding tissue and therefore proper target gene induction. In addition, Lipophorin remains to be the major lipid source for the cells of wing imaginal disc, required for their membrane maintenance and other purposes. This is especially important in regard of the critical implications of lipids in regulation of Hedgehog signaling – the repression of Smoothened by Patched. Thus, the role of Lipophorin might incorporate more than a mere function as a morphogen carrier. It is an intriguing possibility that Lipophorin might carry specific lipids, which can be utilized by Patched to provide an efficient Smoothened inhibition.

SCOPE

4 SCOPE OF THE THESIS

The major focus of the present work is to address the mechanistic *nature of Smoothened repression by Patched*. It has been highlighted in the *Introduction* that Patched has been suggested to utilize a specific lipophilic molecule for Smoothened repression. It has also been hypothesized that the SSD of Patched might play a major role in this process by regulating lipid trafficking. Moreover, Lipophorin – the major lipid transporter in *Drosophila* – has been brought into correlation with Hedgehog transport by an earlier work of our laboratory (Panáková et al., 2005). Based on these hints, an interesting and likely possibility exists that Lipophorin might also play an important role in Hedgehog signaling and carry specific lipids, which can be utilized by Patched to provide an efficient Smoothened inhibition.

Thus, first, I would like to determine whether alteration of Lipophorin levels can in any way affect Smoothened activity. Second, if this were the case, I would like to elucidate the functional requirement of Lipophorin for Patched-mediated Smoothened repression. One of the previous studies has indicated that Patched might normally influence Lipophorin trafficking (Callejo et al., 2008). I wish to further investigate the mechanism of this interaction and test whether Patched could in any way influence the trafficking of Lipophorin-derived lipids. If this action of Patched on Lipophorin lipids can be established, I aim to understand whether this function correlates with the negative regulation of Smoothened localization and/or activity. Finally, if this hypothesis can be confirmed, I would like to determine the identity of the inhibitory compound derived from Lipophorin and utilized by Patched for Smoothened repression.

RESULTS

5 RESULTS

As extensively discussed in the *Introduction*, the nutritional function is only one aspect of the roles of Lipophorin in *Drosophila melanogaster*. Previous studies have already linked Lipophorin to the signaling pathway of Hedgehog, showing that Lipophorin serves as a vehicle for the long-range Hedgehog trafficking and is therefore required for the activation of long-range Hedgehog target genes (Panáková et al., 2005).

5.1 Lipophorin is required to reduce Smoothened accumulation on the basolateral membrane

In order to further dissect the functions of Lipophorin in Hedgehog signaling, I investigated how loss of Lipophorin affects Hedgehog signal transduction in the wing imaginal disc. Lipophorin is produced in the fat body of *Drosophila* and reaches the peripheral organs, including the wing imaginal disc, via hemolymph. Therefore, lowering systemic levels of Lipophorin mediated by fat body-driven RNA interference (RNAi) leads to the lack of Lipophorin in the wing disc tissue (Panáková et al., 2005). Thus, utilizing this RNAi construct, I examined the levels, localization and activation status of the individual components of Hedgehog signal transduction machinery in the wing disc.

Smoothened, the central signal transductor of the Hedgehog pathway, is expressed uniformly throughout the wing pouch. Regulation of its activity correlates with its post-transcriptional stabilization, which reflects the levels of Smoothened protein at the basolateral membrane (Denef et al., 2000). Thus, the activation status of Smoothened can be monitored by Smoothened protein levels on the basolateral membrane of the cell. In the wild type wing disc, basolateral Smoothened levels are reduced by the repressive action of Patched in the cells of the anterior compartment.

Basolateral Smoothened accumulates to a high level only in the posterior compartment, where Patched is not expressed and in the anterior compartment cells near the A/P boundary, where higher levels of Hedgehog inhibit the repressive activity of Patched (Denef et al., 2000). Given that Lipophorin facilitates the transport of Hedgehog to the responsive tissue, one may expect that reduction of Lipophorin would mimic the absence of Hedgehog in these cells and induce a loss-of-function signaling phenotype, downregulating the pathway components and consequently repressing the target genes.

Surprisingly, however, as I examined basolateral Smoothened accumulation in wild type wing discs and wing discs from Lipophorin RNAi larvae, I discovered that Smoothened in fact accumulated throughout the anterior compartment of the wing disc (**Fig. 5.1** A-C).

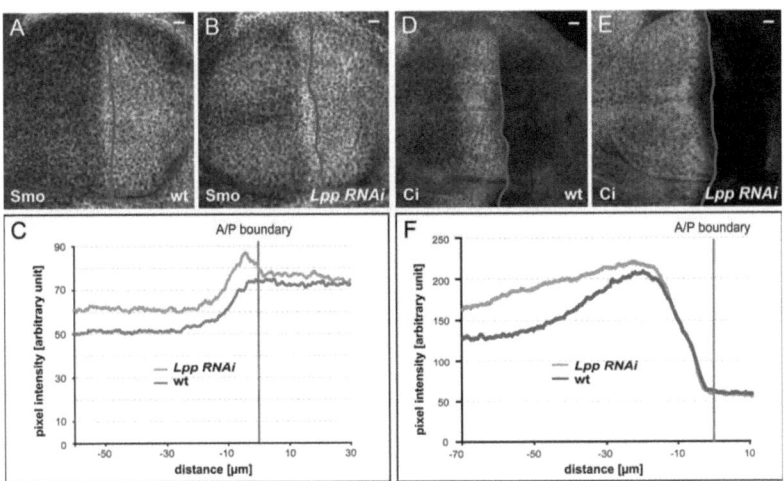

Figure 5.1 **Lipophorin RNAi increases basolateral Smoothened accumulation.**
(A, B) Smo staining in the basolateral region (2.1–4.8 µm from apical surface) of a wild type (A) and LppRNAi (B) wing disc. Smo levels are elevated in the anterior compartment of the LppRNAi disc. (C) Quantification of Smo staining from 19 wild type and 10 LppRNAi discs. Smo staining is

elevated by LppRNAi in the anterior compartment (p<0.0006 for anterior, p<0.4908 for posterior). (D, E) Wing discs from a wild type and a LppRNAi animal stained for Ci_{155}. The range of Ci_{155} stabilization is extended. (F) Quantification of Ci_{155} staining intensity in at least 14 wild type and 14 LppRNAi discs (p<0.0059). The experiment depicted by (D-F) has been performed by Daniela Panáková.
Scale bars = 10 μm. A/P boundaries are indicated by blue lines.

The fact that Lipophorin RNAi resulted in Smoothened accumulation rather than induced its degradation let us suggest that Lipophorin has another, inhibitory, impact on the Hedgehog signaling pathway, which is distinct from its function as a morphogen vehicle. This idea goes in line with the previous finding that Lipophorin RNAi induced an increase of anterior levels of Ci_{155}, the transcription factor in Hedgehog signaling (**Fig. 5.1 D-F**). Thus, apart from its function in Hedgehog trafficking, Lipophorin also functions to repress a subset of Hedgehog signaling events when Hedgehog is absent.

5.2   Lipophorin particles directly affect Smoothened accumulation

How can Lipophorin reduce Smoothened levels at the basolateral membrane? As mentioned in the Introduction, lipoprotein particles play a crucial role in the nutritional lipid transport in *Drosophila*. This function is especially important for the delivery of sterols to the tissues, because *Drosophila* cells cannot synthesize sterols and therefore rely on their dietary uptake (Clayton, 1964). Lipoprotein-delivered sterols can then be used to maintain normal membrane properties. Since Smoothened is a transmembrane protein, I asked if the effects of Lipophorin on Smoothened accumulation and subsequent Ci_{155} stabilization are an indirect consequence of changed membrane properties due to lowered sterol levels in Lipophorin RNAi larvae. To test this possibility, I transferred wild type larvae to the sterol-depleted feeding medium for four days. Wild type larvae fed on sterol-depleted medium supplemented with cholesterol served as a control. Then, I compared the levels of Smoothened and Ci_{155} in the wing discs of sterol-depleted and control larvae with those from Lipophorin RNAi animals (**Fig. 5.1**). If the ectopic Smoothened accumulation was an effect of changed membrane properties in Lipophorin RNAi wing disc cells, then Smoothened, and subsequently Ci_{155}, would accumulate to the same extent in the wing disc cells upon sterol depletion. In contrast, neither Smoothened (**Fig. 5.2** A, B, quantified in C), nor Ci_{155} (**Fig. 5.2** D, E, quantified in F) accumulated throughout the anterior compartment of the wing discs from sterol-depleted larvae, although sterol levels, monitored by staining with filipin, were lowered to the same extent in wing disc cells from both lipid depleted and Lipophorin RNAi larvae (**Fig. 5.2** G-I, quantified in J). Thus, the effects of Lipophorin knock-down on Smoothened accumulation and Ci_{155} stabilization are not an indirect consequence of failure to mobilize dietary lipids, because they cannot be mimicked by mere sterol depletion.

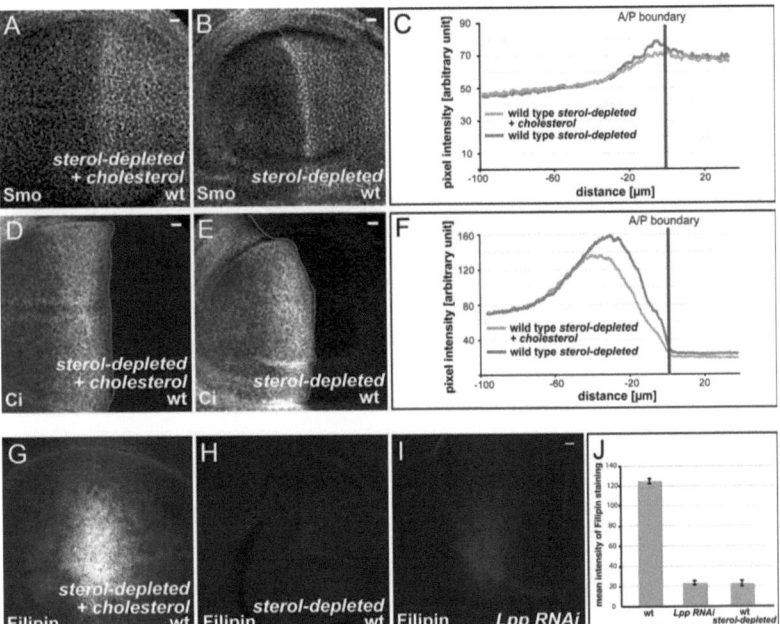

Figure 5.2 **Smoothened and Ci$_{155}$ levels are not affected by dietary lipid depletion.**
(A-C) Smo staining in discs from animals fed on sterol-depleted medium + cholesterol (A) and sterol-depleted medium alone (B), quantified in (C). Smo does not accumulate in the anterior compartment of sterol-depleted discs (p<0.5976) except very near the A/P boundary (p<0.0253). (D-F) Ci$_{155}$ staining of wing discs from wild type larvae that have been transferred to sterol-depleted food (E) or sterol-depleted food supplemented with cholesterol (D) 48 hours after hatching, quantified in (F). Sterol depletion does not change the range of Ci$_{155}$ stabilization. Note that Ci$_{155}$ levels near the A/P boundary are slightly elevated. (G-J) Membrane sterol revealed by filipin staining in wing imaginal discs from larvae with different genotypes or raised under different conditions; each image is a projection of confocal sections through the whole disc. (G) wild type larvae fed on normal food, (H) wild type larvae shifted to sterol-depleted medium 48 hours after hatching, (I) LppRNAi larvae grown on normal food. Stainings from 5 wing discs each are quantified in (J). Membrane sterol levels are equally reduced by LppRNAi and by nutritional sterol depletion.
Scale bars = 10 μm. A/P boundaries are indicated by blue lines.

To further verify that Lipophorin particles acted directly on the wing disc cells, I purified Lipophorin particles from wild type larvae by density gradient centrifugation and incubated them with wing discs explanted from Lipophorin RNAi larvae. Confirming the direct action of Lipophorin on Smoothened, this treatment reversed both basolateral Smo accumulation (**Fig. 5.3** A-D) and Ci$_{155}$ stabilization (**Fig. 5.3** E-H) in the anterior compartment of Lipophorin RNAi wing discs.

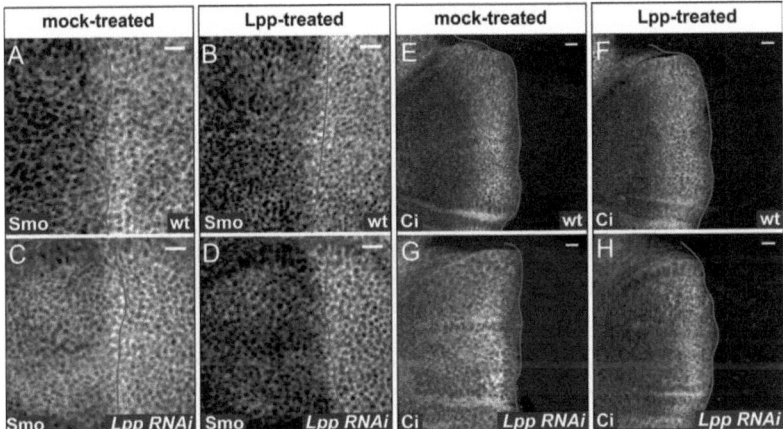

Figure 5.3 **Lipophorin particles reverse Smoothened accumulation and Ci_{155} stabilization.**
(A-D) Smo staining of basolateral sections (2.8 - 4.2 µm below apical surface) of wild type (A, B) and LppRNAi (C, D) wing discs treated with Grace's medium (A, C) or isolated Lpp particles (B, D) for 2 hours. Basolateral Smo accumulation in LppRNAi is reversed by treatment with isolated Lpp particles. (E-H) Ci155 staining of wild type (E, F) and LppRNAi (G, H) wing discs treated with Grace's medium (E, G) or isolated Lpp particles (F, H) for 2 hours. Ci_{155} accumulation in LppRNAi is reversed by treatment with isolated Lpp particles.
Scale bars = 10µm. A/P boundaries are indicated in blue.

5.3 The lipid contents of Lipophorin reduce basolateral Smoothened accumulation and reduce levels of Ci_{155}

Since lipids have been shown to have a great potential in regulation of Smoothened, it was tempting to speculate that the lipid content and not the protein moiety of Lipophorin particles is required to reduce Smooothened and Ci_{155} stability. To test this possibility, I extracted lipids from purified Lipophorin particles, dried them and re-suspended them into protein-free liposomes. The Apolipophorin protein moiety is undetectable in these liposomes (see **Fig. 7.1** in Supplements).

I then incubated discs from Lipophorin RNAi animals with liposomes containing Lipophorin-derived lipids for two hours at a concentration that approximated that of the hemolymph. If inhibition of Smoothened rather required the intact Lipophorin particles, then this treatment should not have any effect on the basolateral Smoothened accumulation in Lipophorin RNAi. However strikingly, addition of Lipophorin-derived lipids was indeed enough to completely reverse the ectopically elevated Smoothened levels in Lipophorin RNAi wing discs to the wild type level (**Fig. 5.4** A-D, quantified in E).

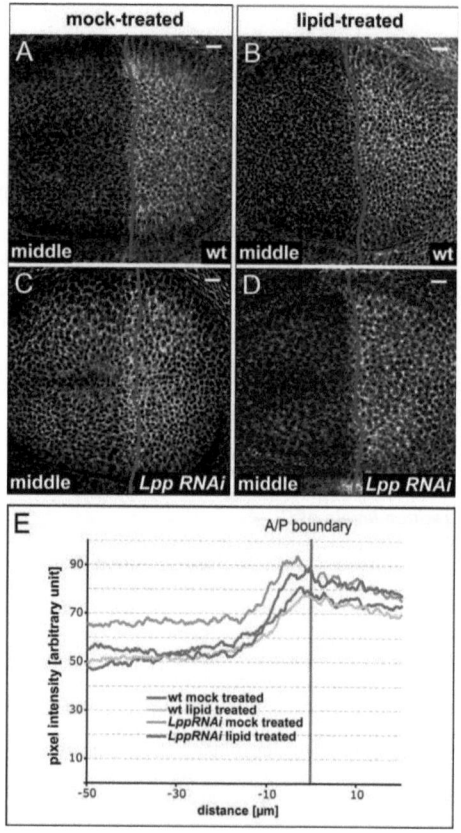

Figure 5.4 **Lipophorin-derived lipids reduce basolateral Smoothened accumulation.**
(A-D) Smo staining in the middle region (2.1–4.8 µm below the apical surface) of a wild type (A, B) or a LppRNAi (C, D) wing disc treated (B, D) or not treated (A, C) with Lpp-derived lipids. Ectopical Smo accumulation is reversed by treatment with Lpp-derived lipids. (E) Quantification of Smo staining of wild type and LppRNAi discs either treated or not treated with Lpp lipids. At least 6 discs were quantified for each condition. Lpp lipids rescue anterior ($p<0.0007$) but not posterior ($p<0.9032$) Smo accumulation in LppRNAi discs. Lpp lipids do not elevate Smo staining intensity in either compartment in wt discs ($p<0.4869$, $p<0.3175$).
Scale bars = 10 µm. A/P boundaries are indicated by blue lines.

Furthermore, this treatment also reversed Ci_{155} accumulation in a subset (27/54) of Lipophorin RNAi wing discs (**Fig. 5.5** A-D).

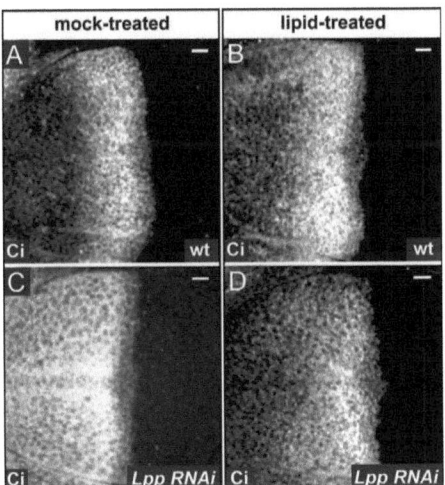

Figure 5.5 **Lipophorin lipids reduce Ci_{155} levels.**
(A-D) Ci_{155} staining of wild type (A, B) and LppRNAi (C, D) wing discs treated with Grace's meidum (A, C) or with Lpp-derived lipids (B, D) for 2 hours. Lpp lipids reduce Ci_{155} accumulation in LppRNAi discs.
Scale bars = 10 µm.

Subsequently, I performed additional controls to confirm the specific effect of Lipophorin-derived lipids on Smoothened accumulation. Thus, I wondered if incubation with liposomes might generally perturb plasma membrane properties and/or protein trafficking in treated cells. Therefore, I examined the levels and localization of another basolateral protein – Arrow (Marois et al., 2006). Incubation with Lipophorin-derived lipids affected neither the levels nor localization of Arrow in treated wing discs, confirming that this treatment does not generally decrease levels of basolateral proteins (**Fig. 5.6** A-D).

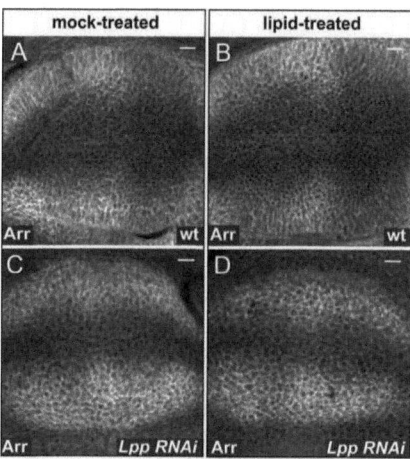

Figure 5.6 **Treatment with Lipophorin lipids does not affect Arrow levels.**
(A-D) Arr staining of basolateral sections (2.8 - 4.2 µm below apical surface) of wild type (A, B) and LppRNAi (C, D) wing discs treated with Grace's medium (A, C) or Lpp-derived lipids (B, D) for 2 hours. Narrowed range of Arr repression in LppRNAi discs is a result of the reduced range of Wingless signaling in LppRNAi (Marois et al., 2006). However, incubation with Lpp lipids has no effect on Arr localization.
Scale bars = 10 µm.

Additionally, I investigated if Smoothened levels are generally sensitive to the bulk lipoprotein lipids. To answer this question, I incubated Lipophorin RNAi wing discs with liposomes containing two of the most abundant Lipophorin lipids, phosphatidylcholine and ergosterol (see **Fig. 7.2** in Supplements). This treatment had no effect on Smoothened accumulation (**Fig. 5.7** A-D) and showed that only specific Lipophorin-derived lipid(s) can reduce levels of Smoothened and consequently destabilize Ci_{155}.

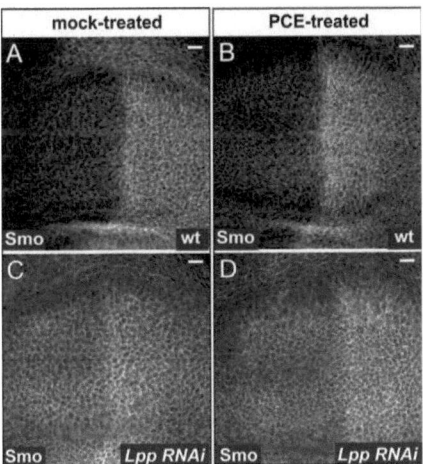

Figure 5.7 **Treatment with phosphatidylcholine or ergosterol does not affect basolateral Smoothened levels.**
(A-D) Smo staining of basolateral sections (2.8-4.2 μm below apical surface) of wild type (A, B) and LppRNAi (C, D) wing discs treated with Grace's medium (A, C) or with liposomes made from phosphatidylcholine + ergosterol (PCE) (B, D) for 2 hours. Treatment with PCE liposomes does not reverse Smo accumulation in LppRNAi discs.
Scale bars = 10μm.

Next, I wondered whether Lipophorin lipids affected Smoothened levels at the level of transcription or whether they influenced Smoothened trafficking by inducing re-localization and thereby depleting Smoothened from the basolateral membrane. Therefore, I examined other subcellular regions of the wing discs treated with liposomes from Lipophorin-derived lipids. Interestingly, the reduction of Smoothened protein levels on the basolateral membrane correlated with the appearance of Smoothened-positive punctate structures in the most apical regions of these cells (**Fig. 5.8** A-H). Colocalization of these Smoothened puncta with internalized red dextran indicated that these vesicles are endosomes and that Lipophorin-derived lipids affected the trafficking of Smoothened protein (**Fig. 5.8** E-G).

Thus, I have demonstrated that the inhibitory effect of Lipophorin on Smoothened is a direct action of its lipid content and not of its protein moiety. Furthermore, I have shown that the action of Lipophorin-derived lipids is specific for Smoothened and affects subcellular Smoothened trafficking, inducing depletion of Smoothened from the basolateral membrane and reduction of its activity, which leads to the destabilization of Ci_{155}. Finally, our observations indicate that the inhibitory effect of Lipophorin on Smoothened is not caused by perturbation of general membrane properties and is not mediated by the bulk lipids transported by Lipophorin and required for membrane maintenance.

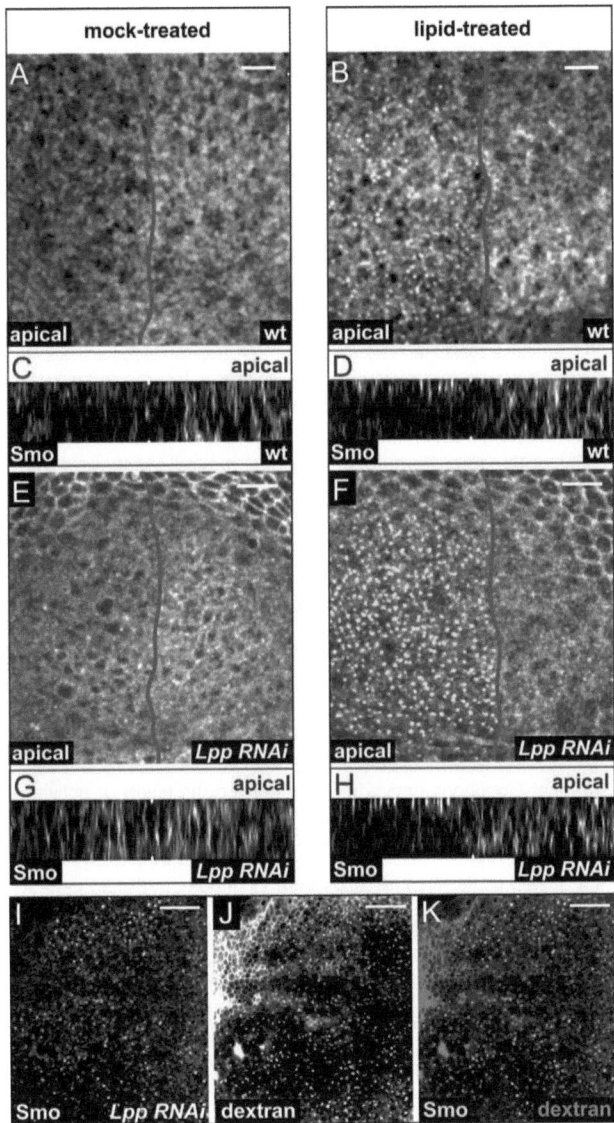

Figure 5.8 **Lipophorin lipids induce translocation of Smoothened from the basolateral membrane to apical endosomes.**
(A-H) Smo staining in the apical region (0.7–2.8 μm below the apical surface) of a wild type (A-D) or a LppRNAi (E-H) wing disc treated with Lpp lipids (B, D, F, H) or mock-treated (A, C, E, G). Upon addition of lipids to LppRNAi wing discs, Smo accumulates in apical punctate structures in the anterior compartment. (C, D, G, H) show z-sections of wing discs from wild type (C, D) or LppRNAi larvae (G, H) treated with Grace's medium (C, G) or with Lpp-derived lipids (D, H). Smo, elevated at the basolateral membrane in LppRNAi discs, translocates to apical puncta upon treatment with Lpp lipids. (I-K) show subapical sections of the anterior compartment of a LppRNAi disc incubated with Lpp lipids and red fluorescent dextran (J and K red) for 2 hours and stained for Smo (I and K green). Smo colocalizes with red dextran in endosomes.
Scale bars = 10 μm. A/P boundaries are indicated by blue lines.

5.4 The Sterol-Sensing Domain of Patched makes Lipophorin-derived lipids available for Smoothened repression

The results of the previous section gave rise to two major questions: first, do Lipophorin-derived lipids repress Smoothened in an interaction with Patched and, if so, what are the requirements for this ensemble? Second, where and how do Lipophorin-derived lipids encounter Smoothened in the cell?

Regarding the first question, I have considered two following scenarios. One possibility was that Lipophorin-derived lipids regulated the levels or activity of Patched protein. Alternatively, regarding the homology of Patched with transmembrane transporters, Patched could mobilize lipids from Lipophorin particles, thus making them "available" to destabilize Smoothened.

In cells mutant for Patched, Smoothened cannot be repressed and consequently accumulates at the basolateral membrane (Strutt et al., 2001). If our first assumption is true, then addition of Lipophorin-derived lipids to these cells would not be expected to reverse ectopic Smoothened accumulation. However, if Patched was merely responsible for lipid mobilization, then direct addition of lipids in liposomes, already extracted and hence mobilized from Lipophorin particles, might reverse basolateral Smoothened accumulation in Patched mutant cells.

I tested these possibilities by applying Lipophorin-derived lipids to the wing discs containing clones of cells totally missing Patched protein – tissue homozygous for $patched^{IIw}$ – and wing discs containing clones of cells homozygous for $patched^{SSD}$. $Patched^{SSD}$ harbors a point mutation in the SSD, which was shown to be crucial for Smoothened repression and hypothesized to be responsible for transporter function of Patched (Martin et al., 2001; Strutt et al., 2001).

Lipophorin-derived lipids did not reduce Smoothened accumulation in tissue totally missing Patched protein (**Fig. 5.9** A-D, quantified in I). However, interestingly, they did reduce Smoothened accumulation in tissue homozygous for $patched^{SSD}$, suggesting that the function of Patched SSD can be circumvented by free lipid addition (**Fig. 5.9** E-H, quantified in J). Thus, these results suggest that Lipophorin-derived lipids cannot act in the total absence of Patched protein. Nevertheless, they show that already mobilized lipids can substitute for the requirement of Patched SSD, which indeed might function to make lipids within Lipophorin particles available to regulate Smoothened trafficking.

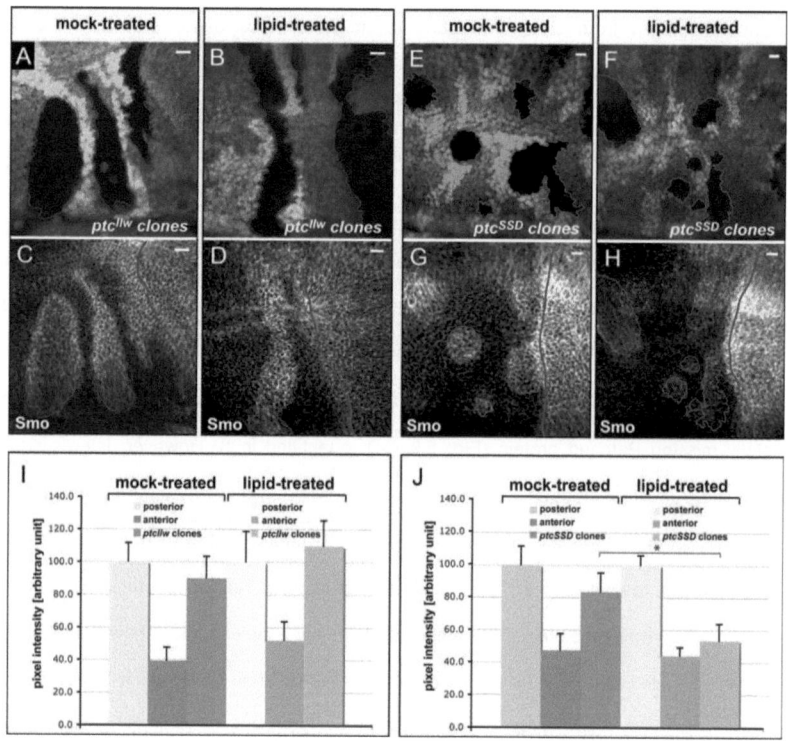

Figure 5.9 **Lipophorin acts with Patched to influence Smoothened trafficking.**
(A-H) Basolateral region of wing discs harboring either or ptc^{llw} (A-D) ptc^{SSD} (E-H) mutant clones indicated by loss of GFP in (A, B, E, F) either treated with Lpp lipids (B, D, F, H) or mock treated (A, C, E, G) stained for Smo (C, D, G, H). ptc^{llw} is an amorphic allele harboring a nonsense mutation at codon 43. ptc^{SSD} is a missense mutation substituting Asn for Asp at position 583 in the sterol sensing domain (the identical mutation contained in the Ptc^{SSD} transgene), it also contains a silent Val to Met substitution at position 1392 (Martin et al., 2001). (I, J) Quantification of Smo staining intensity in discs harboring ptc^{llw} (I) or ptc^{SSD} (J) clones either treated or not treated with Lpp lipids. At least 6 clones were quantified for each condition. Yellow bars: staining intensity in posterior compartment. Red bars: staining intensity in anterior compartment (not the A/P boundary). Green bars: staining intensity in clones. Smo staining intensity is not significantly reduced by Lpp lipids in ptc^{llw} clones, but is reduced in ptc^{SSD} clones (p=$1.6*10^{-13}$).
Scale bars = 10 µm. A/P boundaries are indicated by blue lines.

5.5 Patched induces accumulation of Lipophorin in early endosomes

Given that Lipophorin-derived lipids act in conjunction with Patched to inhibit Smoothened, I asked how Patched might gain access to Lipophorin particles in the first place. Lipophorin particles are internalized and enter the endocytic pathway (Tufail and Takeda, 2009). Thus, I wondered if Patched encountered Lipophorin in endosomes. Therefore, I over-expressed Patched in the dorsal compartment of the wing disc and examined the localization of Lipophorin. I found that Lipophorin strongly accumulated in the Patched-overexpressing compartment (**Fig. 5.10** A, B), which has also been observed elsewhere (Callejo et al., 2008). Interestingly, I additionally discovered that Patched and Lipophorin colocalized in apical punctate structures, which revealed to be Rab5-positive early endosomes (**Fig. 5.10** D-F). Whereas it is possible that general perturbation of the endocytic pathway might impair the progression from early to late endosomes (Rink et al., 2005), I did not observe any alteration in the size or number of either Rab5 or Rab7 positive endosomes, suggesting that Patched over-expression did not produce such artifacts (**Fig. 7.3** in Supplements). Thus, I have concluded that Patched specifically induced Lipophorin accumulation in Patched-positive early endocytic compartment.

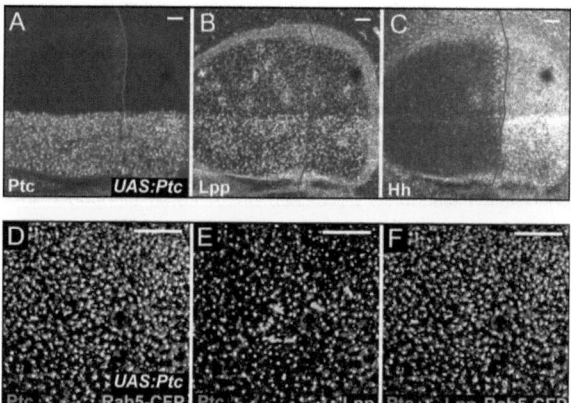

Figure 5.10   **Patched induces Lipophorin accumulation in Rab5-positive endosomes.**
(A-C) apical region (to 0.7–2.8 μm below apical surface) of a wing disc over-expressing Ptc in the dorsal compartment for 18h, stained for Ptc (A), Lpp (B) and Hh (C). Ptc over-expression causes Lpp accumulation independently of Hh (the ratio of average Lpp staining intensity in the dorsal and ventral (Lpp_D/Lpp_V) compartments is 1.5). To compare the endogenous Ptc expression, see (**Fig. 5.13** C). (D-F) apical section (0.7 - 2.1 μm) of a single wing disc ubiquitously expressing low levels of Rab5-CFP that over-expresses Ptc in the dorsal compartment imaged for Ptc (D-F, red), CFP-Rab5 (D and G, green) and Lpp (E, green and F, blue). The dorsal compartment is shown. Lpp and Ptc are found in Rab5-positive endosomes (in the dorsal compartment, 85.9 % of Ptc colocalizes with Lpp, 84.7 % of Ptc colocalizes with Rab5; 79.3 % of Lpp colocalizes with Ptc, 70.3 % of Lpp colocalizes with Rab5; 44.9 % of Rab5 colocalizes with Ptc, 43.4 % of Rab5 colocalizes with Lpp; (P-Value)Costes = 1.0).
Scale bars = 10μm. A/P boundaries are indicated by blue lines.

5.6   The structural requirements of Patched

To investigate the structural requirements of Patched for Lipophorin recruitment, I over-expressed different mutant forms of Patched in the dorsal compartment (to compare expression levels, see **Fig. 7.4** in Supplements).

Patched[1130X] is a mutant form of Patched protein, which misses the C-terminal tail. This region has been shown to be required for Smoothened repression, as well as for Patched internalization and turnover (Lu et al., 2006). I found that this mutant was no longer able to recruit Lipophorin when over-expressed (**Fig. 5.11** A-C). In contrast, Patched^{SSD} with a mutated SSD still was able to recruit Lipophorin (**Fig. 5.11** D-F).

Since Patched is the receptor for Hedgehog, which can be carried on Lipophorin particles, I asked if Patched recruited Lipophorin via Hedgehog. Therefore, I over-expressed Patched$^{\Delta loop2}$, which is missing the sequences required for Hedgehog binding (Marigo et al., 1996), in the dorsal compartment of the wing disc. Subsequently examining Lipophorin localization, I discovered that this Patched mutant form nevertheless efficiently recruited Lipophorin to Patched-positive endosomes (**Fig. 5.11** G-I). Additionally, Patched recruits Lipophorin in the total absence of Hedgehog, as seen in the wing discs from hh^{ts}/hh^{ts} larvae ((Khaliullina et al., 2009), supplementary material).

Thus, Patched affects Lipophorin trafficking independently of Hedgehog and requires its C-terminal tail for this purpose.

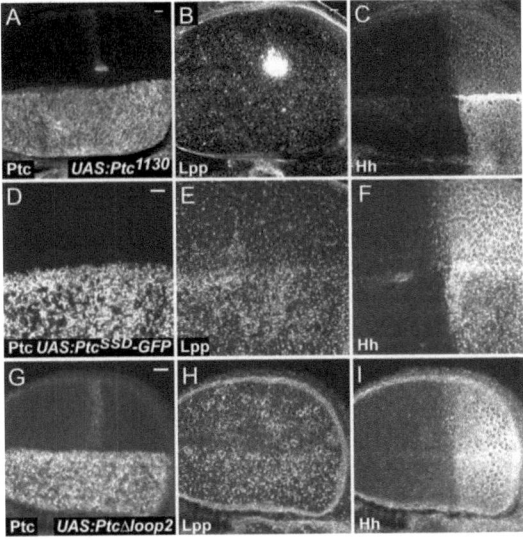

Figure 5.11 **The C-terminal region of Patched is essential to recruit Lipophorin.**
(A-C) a wing disc over-expressing the mutant Ptc^{1130}, which is not well internalized (Lu et al., 2006), in the dorsal compartment for 18h, stained for Ptc (A), Lpp (B) and Hh (C). Ptc^{1130} does not cause Lpp accumulation (the ratio of average Lpp staining intensity in the dorsal and ventral (Lpp$_D$/Lpp$_V$) compartments is 1.0). (D-F) a wing disc over-expressing the mutant Ptc^{SSD}-GFP in the dorsal compartment for 18h, stained for Ptc (D), Lpp (E) and Hh (F). Mutating the SSD does not prevent Lpp accumulation (the ratio of average Lpp staining intensity in the dorsal and ventral (Lpp$_D$/Lpp$_V$) compartments is 1.6). (G-I) a wing disc over-expressing the mutant Ptc$^{\Delta loop2}$ in the dorsal compartment for 18h, stained for Ptc (G), Lpp (H) and Hh (I). Although it cannot bind Hh, Ptc$^{\Delta loop2}$ recruits Lpp (the ratio of average Lpp staining intensity in the dorsal and ventral (Lpp$_D$/Lpp$_V$) compartments is 1.4).
Scale bars = 10 μm.

5.7 *Patched diverts a subset of internalized Lipophorin to Patched-positive endosomes*

How exactly did Patched cause Lipophorin accumulation in the endosomes? It can happen by two mechanisms: either Patched increases Lipophorin uptake or it promotes its longer half-life and decreases its degradation. To distinguish between these two possibilities, I established an uptake assay with labeled Lipophorin. Purified Lipophorin particles were labeled with fluorescent dye Alexa546 and applied to explanted wing discs over-expressing Patched in the dorsal compartment. For the uptake, wing discs were incubated for 10 minutes at 22°C with labeled Lipophorin, then washed, fixed and examined the localization of the Alexa546-labeled Lipophorin. After this incubation, labeled Lipophorin particles were present in endosomes throughout the wing disc and were equally abundant in the dorsal and ventral compartments (**Fig. 5.12 A-C**). Thus, while Patched and Alexa546-Lipophorin are rapidly incorporated into the same endosomes, Patched over-expression did not increase the rate of Alexa546-Lipophorin uptake. To test the general sensitivity of the assay, I applied labeled Lipophorin to explanted wing discs over-expressing LDL receptor homologue LpR1-GFP

((Khaliullina et al., 2009), supplementary material) in the dorsal compartment, then washed, fixed and examined the localization of the Alexa546-labeled Lipophorin. I observed that LpR1-GFP increased the rate of Lipophorin uptake, as was expected for a lipoprotein receptor (**Fig. 7.5** A, B in Supplements), indicating that this assay was sensitive to such changes.

To address the further fate of Lipophorin after its uptake, I applied Alexa546-Lipophorin to explanted wing discs over-expressing Patched in the dorsal compartment, washed and incubated the wing discs for a further 20-40 minutes in Lipophorin-free medium, then fixed and examined Alexa546-Lipophorin localization. The results showed that Lipophorin particles continuously disappeared from the ventral compartment cells but were still visible for up to 20 minutes in Patched over-expressing cells, resembling the situation in the steady state (compare **Fig. 5.10** A-C with **Fig. 5.12** D-F). Thereby, Lipophorin also accumulated with Patched in a Rab5-positive early endocytic compartment (**Fig. 7.6** in Supplements). By 40 minutes, Lipophorin was mostly degraded in both compartments (**Fig. 5.12** G-I). These results showed that Patched decreased the rate of Lipophorin degradation after internalization. Consistent with that, Lipophorin was degraded normally and did not accumulate in cells over-expressing LpR1-GFP (**Fig. 7.5** C, D in Supplements).

Thus, these data have shown that Patched did not increase Lipophorin uptake, being unlikely an endocytic receptor for Lipophorin. Rather, the results suggested that the degradation of Lipophorin was retarded as soon as it reached Patched-positive early endosomes.

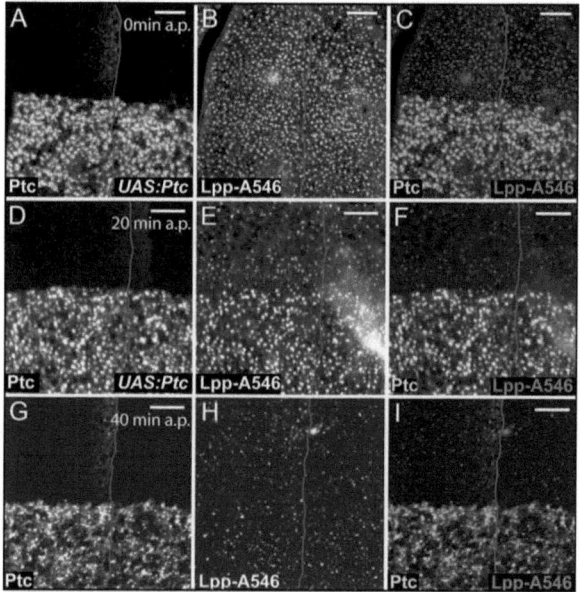

Figure 5.12 **Patched does not affect Lipophorin internalization, but decreases its degradation.**

(A-C) show internalization of purified and fluorescently labeled Lpp (Lpp-Alexa546) by a wing disc over-expressing Ptc in the dorsal compartment. The disc was incubated for 10 minutes with Lpp-Alexa546, then washed and fixed immediately (0min after pulse = 0a.p.). Ptc (A and C green) and Lpp (B and C red) colocalize within 10 minutes (64.9 % of Ptc colocalize with Lpp, 73.0 % of Lpp colocalize with Ptc; (P-Value)$_{Costes}$ = 1.0). Lpp internalization does not increase in Ptc over-expressing cells. (D-F) a wing disc over-expressing Ptc (D and F green), after 10 minute incubation with Lpp-Alexa546, wash and further 20 minute incubation in medium alone (20min after pulse = 20 a.p.). Lpp-positive endosomes are still abundant in Ptc over-expressing tissue, whereas they are much faster degraded in wild type tissue (E, F red). Thus, Ptc slows Lpp degradation (the ratio of average Lpp staining intensity in the dorsal and ventral (Lpp$_D$/Lpp$_V$) compartments is 1.5). (G-I) show a wing disc over-expressing Ptc (G and I, green) after a 10 minute incubation with labeled Lpp particles, then washed and incubated in medium alone for 40 minutes (40min after pulse = 40 a.p.). Most Lpp (H and I, red) has been degraded in both compartments (the ratio of average Lpp staining intensity in the dorsal and ventral (Lpp$_D$/Lpp$_V$) compartments is 1.0.

Scale bars = 10µm. A/P boundaries are indicated by blue lines.

I could unambiguously visualize the effect on Lipophorin trafficking upon overexpression of Patched. However, the endogenous Patched present in the anterior compartment is not sufficient to obviously increase Lipophorin accumulation there (note Lipophorin levels in the wild type ventral compartment in **Fig. 5.10** A, B). I hypothesized that Patched might influence only a small fraction of Lipophorin internalized in the wing disc tissue. This would not be surprising given the important nutritional function of Lipophorin, mentioned above. In fact, in the wing disc Lipophorin can be internalized by a large number of different receptors with the potential to contribute to Lipophorin trafficking in the wing disc cells (Khaliullina et al., 2009).

In order to dissect the subset of Lipophorin affected by the endogenous Patched, I exploited the fact that Lipophorin in Patched-positive endosomes should be degraded more slowly than most Lipophorin internalized by disc cells. To specifically visualize Lipophorin that was degraded more slowly than on average, I incubated explanted wing discs in Lipophorin-free medium for 2 hours. During this time period, most Lipophorin that was internalized *in vivo* should pass through the degradative pathway and disappear, revealing the tissue distribution of any Lipophorin subpopulation with a longer half-life. Indeed, examining Lipophorin localization in those wing discs after a 2-hour incubation reveals a stable population of Lipophorin that is found specifically in the anterior compartment, where Patched is expressed (**Fig. 5.13**). These data support the idea that, whereas most Lipophorin is internalized and degraded rapidly, Patched redirects trafficking of a small subset of Lipophorin particles to a more stable endocytic compartment.

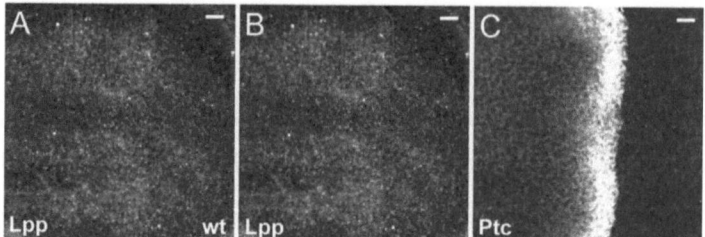

Figure 5.13 **Patched diverts trafficking of a subset of internalized Lipophorin.**
(A-C) apical region (0.7–2.8 μm below apical surface) of a wild type wing disc incubated in Grace's medium for 2 hours and stained for Lpp (A, B) and Ptc (C). The ratio of average staining intensity in the posterior and anterior (Lpp_P/Lpp_A) compartments is 1.2. Lpp is retained preferentially in the anterior compartment cells where Ptc is expressed.
Scale bars = 10 μm. A/P boundaries are indicated by blue lines.

5.8 Mutation of the Patched Sterol-Sensing Domain perturbs lipid trafficking from Patched-positive endosomes

Patched is homologous to bacterial transmembrane transporters and NPC-1, having in common the SSD (Tseng et al., 1999). The SSD of NPC-1 is required for efflux of sterols, sphingolipids and other lipids from late endosomes. These include lipids derived from internalized lipoprotein particles (Wojtanik and Liscum, 2003; Ikonen and Holtta-Vuori, 2004; Mukherjee and Maxfield, 2004).

Since Patched was able to stabilize Lipophorin in endosomes, I hypothesized that sequestration of Lipophorin by Patched in these endosomes might give Patched access to Lipophorin-derived lipids. I then asked whether, in homology to NPC-1, the SSD of Patched similarly affected trafficking of Lipophorin-derived lipids from Patched-positive endosomes. As mentioned above, Drosophila tissues are auxotroph for sterols and rely on their delivery by lipoproteins. Thus, visualization of sterols by Filipin, which stains free unesterified sterols, would be a convenient marker for Lipophorin-derived lipids. I over-expressed wild type Patched, Patched^{SSD}-GFP and Patched^{SSD}, both with mutated SSDs, and Patched[1130] with an intact SSD but missing its C-terminal tail, in the dorsal compartment of the wing disc, stained the tissue with Filipin and examined sterol distribution.

I found, that over-expression of wild type Patched did not perturb sterol distribution (compare **Fig. 5.14** A with **Fig. 5.14** B). In contrast, both Patched[SSD]-GFP and Patched[SSD] caused sterol accumulation in Patched-positive endosomes at the apical region of the wing disc cells (**Fig. 5.14** D-H). Interestingly, Patched[1130] over-expression did not cause endosomal sterol accumulation (**Fig. 5.14** C). This result confirmed our assumption that the SSD of Patched is specifically required for the exit of sterols from Patched-positive endosomes. Moreover, it showed that the ability to affect sterol trafficking correlates with the capability of Patched protein to sequester Lipophorin, because Patched and Patched[SSD]-GFP, but not Patched[1130] possess the ability to stabilize Lipophorin in Patched-positive endosomes.

Combining the structural requirements of Patched for Lipophorin recruitment, I have concluded that Patched stabilized Lipophorin in the Patched-positive endosomes, and regulated the sterol efflux from these endosomes by its SSD.

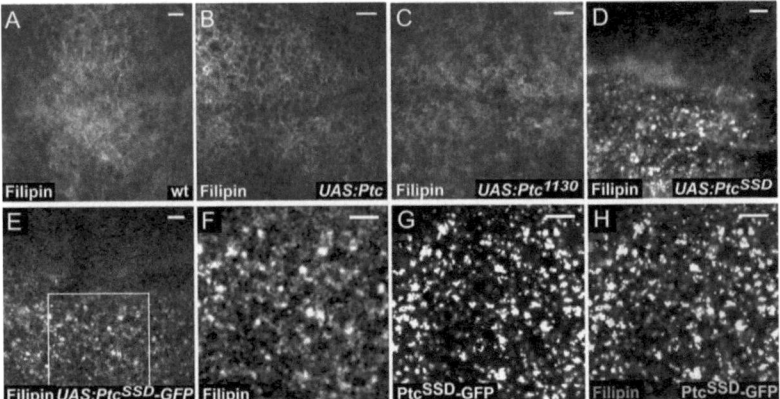

Figure 5.14 **Mutation of the Patched Sterol-Sensing Domain prevents endosomal sterol efflux.**
(A) sub-apical region of wild type wing disc stained with Filipin. (B-H) sub-apical region of discs expressing following different Ptc variants in the dorsal compartment: wild type Ptc (B), Ptc^{1130} (C), Ptc^{SSD} (D), Ptc^{SSD}-GFP (E-H). (F-H) magnified images of the region boxed in (E): Filipin staining (F, H red) and Ptc^{SSD}-GFP (G and H green). Ptc^{SSD} and Ptc^{SSD}-GFP cause sterol accumulation in Ptc endosomes, whereas wild type Ptc and Ptc^{1130} do not (68.7 % of Ptc^{SSD}-GFP colocalize with Filipin, 88.9% of Filipin colocalize with Ptc^{SSD}-GFP; (P-Value)$_{Costes}$ = 1.0).
Scale bars = 10 µm.

5.9 Lipids mobilized by Patched are derived from newly delivered Lipophorin

Next, I asked whether the sterol that accumulated in Patched^{SSD}-positive endosomes is derived from Lipophorin, which has been sequestered in these endosomes by Patched. Although it was a tempting hypothesis, it was also possible that membrane sterol, previously delivered by Lipophorin, would also be trapped in endosomes by mutation of the SSD of Patched. How could one specifically mark the sterol, newly delivered by Lipophorin and track its cellular fate upon the uptake? For this purpose, I made use of BODIPY-cholesterol (Holtta-Vuori et al., 2008), which I fed to the larvae coincident with induction of either Patched or Patched^{SSD} over-expression in the dorsal compartment of their wing discs. After 48 hours, I monitored the sterol distribution in these wing discs by live imaging for BODIPY. Confirming our previous idea, BODIPY-cholesterol accumulated in apical endosomes of cells over-expressing Patched^{SSD} but not wild type Patched (**Fig. 5.15**). Thus, at least part of the accumulated sterol has been delivered to discs subsequent to the induction of Patched expression.

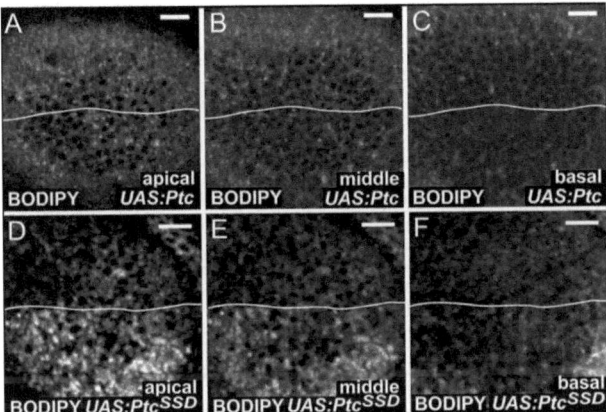

Figure 5.15 **The specific effect of the Sterol-Sensing Domain of Patched on newly delivered sterol.**
(A-F) wing discs from larvae, which have been transferred from normal food to lipid-depleted medium containing 6,5μg/ml BODIPY-cholesterol coincident with induction of Ptc (A-C) or Ptc^{SSD} (D-F) over-expression in the dorsal compartment. (A, D) show apical, (B, E) – middle and (C, F) – basal sections of the wing disc (0.7 - 2.1 μm, 2.8 - 4.2 μm and 4.8 - 6.3 μm below apical surface, respectively). Newly delivered BODIPY-cholesterol accumulates in cells over-expressing Ptc^{SSD} but not Ptc.
Scale bars = 10 μm.

Further, it was tempting to assume that the stabilization of Lipophorin in Patched-positive endosomes is required for its sequestration by the SSD of Patched. To test this hypothesis, I labeled the protein moiety of the purified Lipophorin with Alexa546, whereas its lipid contents were marked by BODIPY-cholesterol. Next, wing discs over-expressing Patched^{SSD} in the dorsal compartment were incubated with these particles for 40 minutes and subjected to live imaging. Whereas Alexa546-labeled protein moiety of the Lipophorin particles was degraded after 40 minutes, consistent with our previous results (see **Fig. 5.12 G-I**), BODIPY-cholesterol-labeled Lipophorin cargo accumulated only in cells of the Patched^{SSD}-over-expressing compartment (**Fig. 7.7** in Supplements).

This showed that the SSD of Patched indeed acted to derive lipids from the delivered Lipophorin particles and that trafficking of at least one Lipophorin lipid – sterol – is perturbed when the function of Patched SSD is impaired.

5.10 Mutation of the Patched Sterol-Sensing Domain perturbs Smoothened trafficking from Patched-positive endosomes

At the end of section *5.3*, the following questions were asked: first, do Lipophorin-derived lipids repress Smoothened in an interaction with Patched and, if so, what are the requirements for this ensemble? Second, where and how do Lipophorin-derived lipids encounter Smoothened in the cell?

Regarding the first question, my investigations have led to the conclusion that, after its internalization, a small subset of Lipophorin particles is stabilized in endosomes by Patched, the main requirement for this process being the C-terminal tail of Patched. Consequently, Lipophorin particles are sequestered by the SSD of Patched, which mobilizes Lipophorin-derived lipids – sterols at the least.

However, it still remained unclear where and how could mobilized Lipophorin-derived lipids encounter and affect Smoothened.

Based on many studies investigating the regulation of Smoothened by lipophilic compounds (Frank-Kamenetsky et al., 2002; Chen et al., 2002a; Chen et al., 2002b), I hypothesized that Lipophorin-derived lipids would require close proximity to exert their effect on Smoothened trafficking. Thus, it was tempting to speculate that Lipophorin sequestration by Patched and subsequent Smoothened inhibition took place in the same endosomal compartment. Since mobilization of Lipophorin-derived lipids depended on the SSD, I wondered if mutation of this domain also affected Smoothened trafficking. Thus, I compared the distribution of Smoothened in cells of the wing discs over-expressing either wild type Patched or Patched^{SSD} in the dorsal compartment. I observed that, in striking contrast to the wild type form, Patched^{SSD} expression caused a dramatic accumulation of Smoothened in apical endosomes together with Patched^{SSD} and Lipophorin (compare **Fig. 5.16** A-H with **Fig. 7.8** A in Supplements). In the anterior compartment, Patched^{SSD} expression elevated basolateral Smo levels (**Fig. 5.16** I-L), which was expected due to the activation of Smoothened signaling by this mutant form reported previously (Martin et al., 2001).

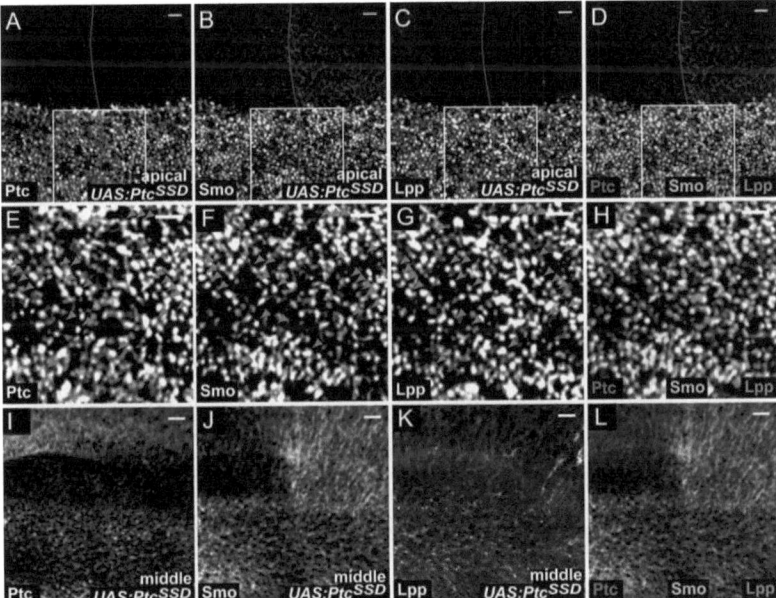

Figure 5.16 **Mutation of the Patched Sterol-Sensing Domain traps Smoothened in Patched-positive endosomes.**

(A-D) apical section (0.7 - 2.1 μm below apical surface) of a wing disc, which has been over-expressing Ptc^{SSD} in the dorsal compartment for 18h, stained for Ptc (A, E and D, H blue), Smo (B, F and D, H green) and Lpp (C, G and D, H red). (E-H) are magnified images of the region boxed in (A-D). Ptc^{SSD} over-expression causes accumulation of Smo in Ptc- and Lpp-positive apical endosomes. Red arrowheads indicate examples of triple colocalization. In the dorsal compartment, 94.4 % of Ptc colocalize with Lpp; 94.8 % of Ptc colocalize with Smo; 91.6 % of

Lpp colocalize with Ptc; 96.7 % of Lpp colocalize with Smo; 81.8 % of Smo colocalize with Ptc; 85.9 % of Smo colocalize with Lpp; (P-Values)$_{Costes}$ = 1.0). (I-L) basolateral region of a wing disc over-expressing Ptc^{SSD} in the dorsal compartment for 18h, stained for Ptc (I and L blue), Smo (J and L green) and Lpp (K and L red). Over-expression of Ptc^{SSD} elevates basolateral Smo. Scale bars = 10 μm. A/P boundaries are indicated by blue lines.

To ask whether Patched^{SSD} might generally perturb the trafficking of basolateral membrane proteins, I examined the apical localization of Fasciclin III and Arrow in Patched^{SSD} over-expressing cells. Unlike Smoothened, neither Fasciclin III nor Arrow accumulated in endosomes with Patched^{SSD}, nor were their levels on the basolateral membrane altered (**Fig. 5.17**). Thus, mutation of the SSD specifically affected trafficking of both Lipophorin-derived lipids and Smoothened without perturbing localization of other basolateral membrane proteins.

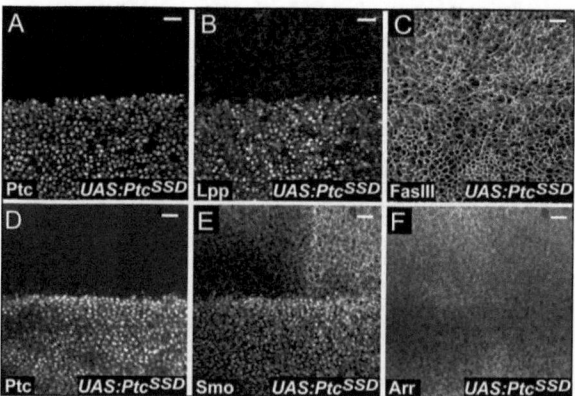

Figure 5.17 **Over-expression of Patched^{SSD} does not affect basolateral proteins FasciclinIII and Arrow.**
(A-F) apical region (0.7 – 2.8 μm below apical surface) of wing discs over-expressing Ptc^{SSD}, which have been stained for Ptc (A, D), Lpp (B), FasIII (C), Smo (E) and Arr (F). Over-expression of Ptc^{SSD} causes dramatic Smo and Lpp accumulation in apical endosomes, but has no effect on basolateral proteins FasIII and Arr.
Scale bars = 10 μm.

Further, in order to confirm that the SSD of Patched induced endosomal Smoothened accumulation also when expressed at endogenous levels, I examined Smo localization in *patched*^{SSD} mutant cells. I observed a similar, though less dramatic colocalization of Patched^{SSD}, Smoothened and Lipophorin in punctate structures in *patched*^{SSD} mutant tissue (**Fig. 5.18**). Taken together, these observations indicated that Smoothened may normally traffic through Patched-positive endosomes, and that blocking the activity of the SSD of Patched not only causes endosomal lipid accumulation but also alters the trafficking of Smoothened from this compartment.

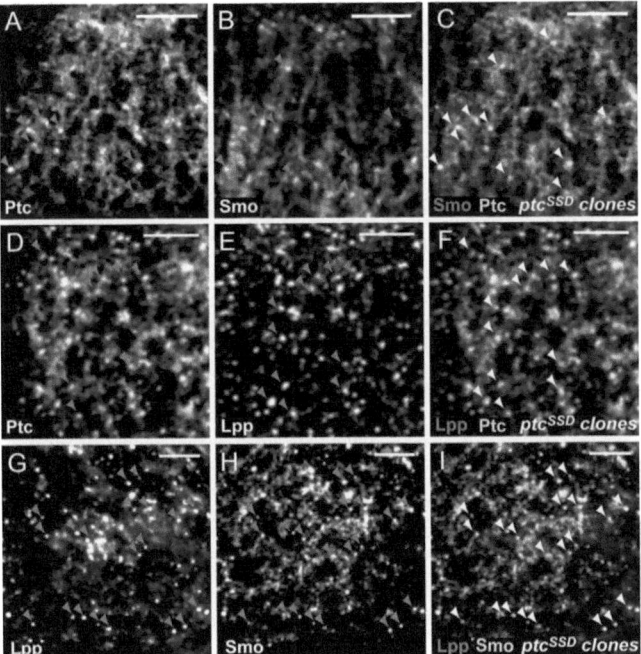

Figure 5.18 **Effects of the mutation in the Sterol-Sensing Domain of Patched on Smoothened trafficking.**

All images show apical confocal sections 0.7–2.8 µm below apical surface. (A-C) a ptc^{SSD} clone stained for Ptc (A and C green) and Smo (B and C red). (D-F) a ptc^{SSD} clone stained for Ptc (D and F green) and Lpp (E and F red). (G-I) a ptc^{SSD} clone stained for Lpp (G and I red) and Smo (H and I green). Smo is found in punctate structures containing Ptc^{SSD} and Lpp. Examples of colocalization are indicated by arrowheads (C, F, I).
Scale bars = 10 µm.

5.11 Purification of the active lipid species from Lipophorin-derived lipids

Having discovered, for a first time, that endogenous lipids regulate Smoothened trafficking and its activity, I was intrigued what the active species might be.

Prior to the purification of the Lipophorin lipid extract, I first sought to determine the active unit of this complex lipid mixture. Therefore, I prepared liposomes with different Lipophorin lipid concentrations, incubated them with wing discs explanted from Lipophorin RNAi larvae and monitored the levels of Smoothened on the basolateral membrane. Testing different lipid concentrations, I found that ca. 350µg/ml total lipid were sufficient to reduce basolateral Smoothened levels in Lipophorin RNAi wing discs to the wild type level (**Fig. 7.8** in Supplements).

Next, I characterized the lipid classes of the Lipophorin lipid mixture by separating them by means of thin layer lipid chromatography (TLC). According to the standards, Lipophorin lipid extract contained neutral lipids, marked by triacylglycerol, phospholipids, marked by phosphoethanolamine and phosphatidylcholine, and sterols marked by cholesterol. However, the extract also contained bands not corresponding to any used standard, which might carry the active species as well. Therefore, visualizing the lipid bands by charring, we scratched all visible bands from the non-charred part of the TLC plate (**Fig. 5.19** A). I prepared liposomes with lipids extracted from the scratched bands, incubated them with Lipophorin RNAi wing discs and monitored the basolateral Smoothened levels. Strikingly, I observed that only lipids from band no. 3

were able to reverse basolateral Smoothened accumulation to the wild type level (**Fig. 5.19** B). Thus, the active species was not a neutral lipid, sterol or phospholipid, since band no. 3 did not correspond to these standards. This was not surprising given the fact that phosphoethanolamine, phosphatidylcholine, diacylglycerol, cholesterol and ergosterol did not show any potential to reduce basolateral Smoothened levels (**Fig. 5.7** and data not shown).

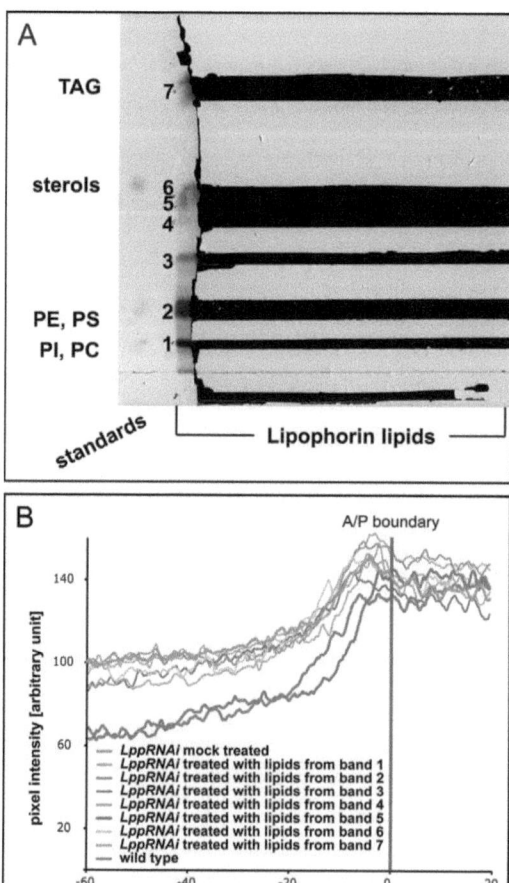

Figure 5.19 Specific fraction (from band 3) of Lipophorin lipids reduces basolateral Smoothened accumulation.

(A) Preparative TLC of Lpp lipid extract with indicated standards. Regions corresponding to the lipid bands 1 to 7, stained on the left part of the TLC plate, were scratched from the right part of the TLC plate and extracted from silica. Liposomes were prepared with the lipids from each fraction and applied to the wing discs from LppRNAi larvae for 2 hours. The discs were fixed, stained for Smo and the staining intensity was compared to that of Smo staining of the wild type discs. (B) shows quantification of Smo staining intensity of LppRNAi wing discs treated with each lipid fraction (from bands 1-7) compared to that of mock-treated LppRNAi wing discs and to that of wild type discs. Elevated Smo levels in LppRNAi wing discs are reversed to the wild type levels upon treatment with Lpp-derived lipids from only band no.3 (p<0.0005). At least 6 discs were quantified for each condition.

In parallel to this approach, I aimed to further purify the crude Lipophorin lipid extract by saponification. Saponification is the hydrolysis of an ester bond under basic conditions to form an alcohol and a carboxylate. By this means, lipid species such as triacylglycerides, diacylglycerides, sterol esters and phospholipids were removed from the Lipophorin lipid extract. I tested whether the non-saponifiable lipid fraction still had the activity to reduce basolateral Smoothened levels in explanted wing discs from Lipophorin RNAi larvae. Interestingly, the non-saponifiable lipids retained the full potential to reverse basolateral Smoothened accumulation (**Fig. 20**).

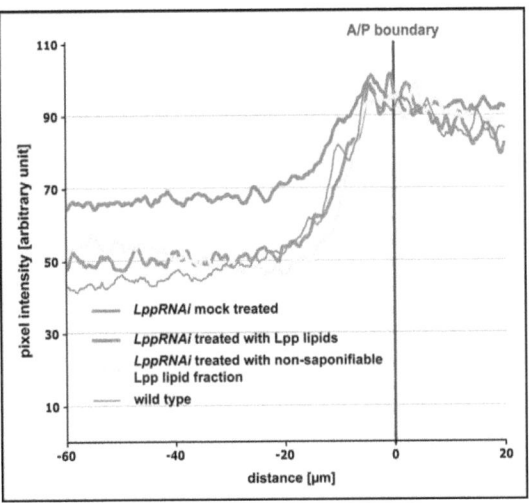

Figure 5.20 Non-saponifiable Lipophorin-derived lipids reduce basolateral Smoothened accumulation.

Quantification of Smo staining of wing discs form LppRNAi larvae treated with Grace's medium, Lpp-derived lipids and the non-saponifiable fraction of Lpp-derived lipids compared to Smo staining intensity of wild type wing discs. Elevated Smo levels in LppRNAi wing discs are reversed to the wild type levels by treatment with Lpp-derived lipids as well as by treatment with only the non-saponifiable fraction of Lpp-derived lipids ($p<0.003$). At least 6 discs were quantified for each condition.

In summary, I have shown here that after internalization, Lipophorin particles are stabilized in an endosomal compartment by Patched, which mobilizes specific Lipophorin lipids via its SSD. This region of Patched also regulates the trafficking of Smoothened from this endosomal compartment, facilitating Smoothened degradation when Lipophorin lipids are present.

DISCUSSION

6 DISCUSSION

6.1 The ongoing mystery of Smoothened repression – clues so far

The mechanism of Smoothened repression by Patched has been one of the most intricate questions ever asked during investigations of the Hedgehog signaling pathway. Numerous studies provided bases for various suggestions concerning the modulation of Smoothened activity. Among these, the most popular hypothesis proposes Smoothened to be negatively regulated by binding of a small lipophilic molecule. Thereby, Patched is thought to regulate availability or presentation of such a molecule to Smoothened. However, there are still some ambiguities left with that theory. For instance, only vertebrate Smoothened has been shown to actually bind lipidic inhibitors so far. Further, while some of these repressor molecules share structural homology with cyclopamine, a sterol derivative, others do not appear to be of any common molecular origin (Frank-Kamenetsky et al., 2002; Chen et al., 2002a; Chen et al., 2002b). Also, it has been generally unclear, where specifically in the cell an interaction between Patched, Smoothened and its inhibitor(s) might occur. Finally, all identified Smoothened inhibitors are not likely to be endogenous molecules.

Additional mode of regulation has been described for the *Drosophila* Smoothened. There, Smoothened activity has mainly been brought into correlation with its subcellular distribution. Thus, stability of Smoothened at the basolateral membrane reflects its activation state (Denef et al., 2000; Nakano et al., 2004). Also, recruitment of Smoothened to the cell surface has been shown to correlate with its increased phosphorylation, which is thought to induce a switch in its conformation (Kalderon, 2005; Zhao et al., 2007). Although this is certainly an attractive model, many steps also remain to be clarified here. For instance, it has not been investigated in detail, what could trigger the step-by-step phosphorylation of Smoothened and its subsequent membrane translocation in response to the Hedgehog signal. Further, the role of Patched in this process remains to be poorly understood.

At first sight, it seems that vertebrates and invertebrates developed two fundamentally different strategies for Smoothened regulation. However, theories mentioned above are not necessarily unconnected and might, in fact, even complement each other. Thus, the link between Smoothened activity and its subcellular localization, well established for the *Drosophila* Smoothened, meanwhile found its confirmation by studies in vertebrates. Thus, activation of mammalian Smoothened results in its recruitment to the primary cilia, which are specialized plasma membrane domains. At the same time, Patched, bound to Hedgehog, translocates to the internal stores. In the absence of Hedgehog, Smoothened is absent from the cilia, whereas Patched is found in this membrane compartment (Corbit et al., 2005; Haycraft et al., 2005; Huangfu and Anderson, 2005; Huangfu and Anderson, 2006; Rohatgi et al., 2007). Thus, combining the previously described data on Smoothened modulation by small lipidic molecules, it has been speculated that in the absence of Hedgehog Patched utilizes an inhibitor to induce a conformational switch and subsequent depletion of Smoothened from the primary cilium. If this view is correct, and the regulation of Smoothened molecules has remained principally similar throughout evolution, it is tempting to speculate that *Drosophila* Smoothened could also be influenced by a lipophilic inhibitor. Thus, this repressor molecule could either directly bind to Smoothened or indirectly regulate its subcellular localization in a Patched-dependent manner.

This is an attractive model, which would be consistent with the established data from both biological systems. Nevertheless, there has as yet been no evidence to suggest that Patched alters lipid trafficking in any way. Furthermore, no such Smoothened-inhibitory molecule has been identified *in vivo* yet. Additionally, if Patched could indeed control the availability of some hydrophobic repressor, it is not clear where this molecule could gain access to Smoothened, since endogenous Patched and Smoothened have not been observed to colocalize in mammalian or invertebrate cells.

The present work provides a strong evidence for an endogenous inhibitory lipid, which is regulating Smoothened activity in *Drosophila*. Furthermore, it sheds some light on the previously unclear mechanistic steps of Patched action on invertebrate Smoothened, which could help to understand the basis of the general regulation of Smoothened molecules.

6.2 The Sterol-Sensing Domain of Patched regulates Smoothened trafficking

One of the main obstacles on the way to establishing a plausible mechanism for Patched-mediated Smoothened repression has been the lack of any evidence of the close proximity of these molecules in the cell. In this study, endogenous Smoothened and Patched have for the first time been observed to colocalize in the same subcellular compartment, the only prerequisite for this colocalization being the dysfunction of Patched SSD (**Fig. 5.16**). The SSD of Patched has been already reported to constitute a crucial structural requirement for Patched ability to inhibit Smoothened signaling (Martin et al., 2001).

What could be the exact function of Patched SSD in this process? A structural homology between Patched and NPC-1 as well as their broader similarity with bacterial RND permeases, which belong to the larger family of multidrug-resistance (MDR) efflux pumps, might provide a substantial clue to this question. The common property of these proteins is their SSD domain, mutation of which interferes with the transporter function in the bacterial transmembrane pumps and with the function of NPC-1 (Watari et al., 1999). Strikingly, it is exactly that mutation in the SSD of Patched, which perturbs Smoothened trafficking and forces it to accumulate in Patched^{SSD}-positive endosomes. This suggests that the direct function of Patched SSD might be similar to a transmembrane transporter, controlling trafficking of a hydrophobic messenger molecule, which could catalytically inhibit Smoothened – an attractive possibility repeatedly suggested during investigations of the Hedgehog signaling pathway (Taipale et al., 2002; Chen et al., 2002b).

How could a lipid influence Smoothened activity? The fact that various small lipidic compounds can inhibit mammalian Smoothened led to a straightforward hypothesis that Patched facilitates binding of such a compound to Smoothened. However, the relationship of Smoothened with the family of seven-transmembrane receptors indicates a much broader range of possibilities for Smoothened regulation by a lipid. Thus, a fundamental property of these receptors is their ability to realign their transmembrane helices in response to hydrophobic cues, which results in a conformational change of the receptor and, therefore, its ability to interact with cytoplasmic effectors (Gimpl et al., 1997; Pitcher et al., 1998). This could lead to an alteration of receptor trafficking and, subsequently, its signaling activity. Thus, increased recycling of the receptor might enhance its signaling, whereas its predominant degradation could result in the neutralization of the signal. Smoothened might have retained this property, since its signaling activity correlates significantly with its subcellular localization. Thus, Smoothened is trafficking and accumulating at the basolateral membrane in the presence of Hedgehog, while being preferentially targeted to the degradative pathway in anterior compartment cells (Denef et al., 2000; Nakano et al., 2004). Notably, the hierarchy of the cell surface localization and activation of Smoothened has not been established so far. Thus, several possible scenarios exist as to how these two processes are linked. One possibility is that Smoothened signaling activity is regulated quantitatively and achieved simply by assembling high levels of Smoothened protein at the basolateral membrane. In this case, basolateral Smoothened trafficking would be required to induce pathway activation. Alternatively, activation of

Smoothened could be accomplished first, by an unknown mechanism, leading to its cell surface translocation, from where Smoothened could then signal. Finally, translocation of Smoothened to the basolateral membrane might be a consequence of relieved repression of Smoothened, whereas its activation would be achieved by another, separate signal.

In general, trafficking of the proteins between different subcellular compartments is guided by numerous effector molecules, which associate with the endosomal membranes (Stenmark, 2009). The recruitment of these effectors is strongly influenced by the endosomal lipid composition, as seen for the different Rab proteins. Thus, lipid accumulation in endosomes of NPC-1 mutant cells strongly influences the activity of Rab7 (Lebrand et al., 2002), Rab9 (Ganley and Pfeffer, 2006) and Rab4 (Choudhury et al., 2004), perturbing degradation and recycling. Interestingly, NPC-1 is the closest homologue of Patched proteins, triggering an exciting possibility that Patched might function in a similar way to repress Smoothened signaling. In particular, Patched could regulate the lipid composition of the endosomes, where it is present, via its SSD. When Smoothened would traffic through this compartment, it could be exposed to lipids that are mobilized by Patched and that bias Smoothened trafficking towards degradation, terminating its signaling.

Interfering with the SSD function would therefore prevent lipid mobilization by Patched and the balance between Smoothened degradation versus recycling would be favored towards the latter, causing a "traffic jam" in Smoothened trafficking. This would result in increased Smoothened accumulation in the endosomes as well as at the basolateral membrane, as observed in our work (Fig. **5.16**). Accumulation of sterol in Patched^{SSD}-containing endosomes further supports this idea and indicates that the SSD of Patched is indeed required to control the lipid efflux from the endosomes (Fig. **5.14** and **5.15**). However, more detailed investigations are necessary to mechanistically define the action of the SSD. Thus, it needs to be established, whether Patched functions as a lipid flippase or, alternatively, facilitates the intercalation of lipids into the endosomal membrane bilayer.

6.3 Endogenous Smoothened inhibitor is derived from Lipophorin particles by Patched

How could Patched possibly gain access to a Smoothened-regulatory lipid? Generally, lipids are delivered to the wing disc cells in Lipophorin particles. Subsequently, Lipophorin is internalized and enters the endocytic route. Normally, the particles are sequestered, making their lipid contents available for further cellular needs. Mainly, lipids are utilized for the nutritional requirements. Thereby, NPC-1 promotes the transport of neutral lipids out of the lysosomal compartment, employing its SSD and interacting with NPC-2 (Infante et al., 2008a; Infante et al., 2008b; Infante et al., 2008c; Kwon et al., 2009). We have shown that Patched analogously regulates lipid trafficking via its SSD. However, this process does not take place in a degradative compartment, since sterols, which accumulate in Patched^{SSD}-endosomes, do not colocalize with lysosomal markers (not shown). This indicates that Patched might influence the trafficking of specific lipids, redirecting them from the common trafficking route, employed by NPC proteins for the delivery of bulk nutritional lipids. If this is the case, then Patched should also influence the trafficking of Lipophorin particles, which carry these lipids. Indeed, this work shows that Patched stabilizes a fraction of Lipophorin particles internalized by the wing disc cells, in an early endosomal compartment (**Fig. 5.10**). Furthermore, we observed that depletion of Lipophorin from the system strongly interferes with the correct regulation of Hedgehog pathway, since Smoothened and, subsequently, the full-length transcription factor Ci_{155}, accumulate in the wing discs from Lipophorin RNAi larvae (**Fig. 5.1**). This is a direct consequence of the absence of a lipidic Smoothened regulator, since lipids derived from the wild-type Lipophorin particles directly affect Smoothened trafficking and signaling, promoting Smoothened depletion from the basolateral membrane and destabilizing Ci_{155} (**Fig. 5.4** and **Fig. 5.5**).

We succeeded to show that Patched is able to regulate lipid trafficking from the endosomes, where it is present. It is therefore reasonable to assume that stabilization of Lipophorin particles in Patched-positive endosomes would serve the sequestration of Lipophorin by Patched in order to make use of the lipid contents of the particles. If this is the case, then perturbation of the Patched SSD function would lead to impaired trafficking of Lipophorin lipids from Patched-positive compartment, which is exactly what we observed – Lipophorin-derived BODIPY-cholesterol accumulates in endosomes with Patched^{SSD} (**Fig. 5.15**). Furthermore, Patched indeed appears to recruit Lipophorin particles from their main utilization route in the cell and promote the efflux of their specific lipid cargo. Thus, when the protein moiety of Lipophorin particles was marked with Alexa546 and their lipidic contents were labeled with BODIPY-cholesterol, the particles were taken up normally. When Lipophorin reached the Patched-positive compartment, it accumulated there for at least 20 minutes (not shown), consistent with our previous results. After two hours, we did not observe the accumulation of Lipophorin protein moiety labeled by Alexa546, indicating its degradation or rapid recycling. However, the BODIPY-labeled lipid cargo accumulated in the endosomes when the function of Patched SSD was impaired (**Fig. 7.7** in Supplements). Thus, this work provides a first evidence for an endogenous inhibitor of Smoothened signaling – it is a lipid derived from lipoprotein particles and utilized by Patched via its SSD.

Notably, although these results support the idea that a sterol or a sterol derivative might account for the activity of Lipophorin lipid extract, it is also possible that Patched SSD regulates trafficking of other lipids, since RND transporters have been shown to possess extremely broad substrate specificities (Piddock, 2006). This possibility will be discussed further.

6.4 Role of Patched in lipid trafficking

Notably, Patched regulates the trafficking of only a small subset of the whole Lipophorin pool (**Fig. 5.13**), providing a plausible mechanism as to how a specific – and possibly not very abundant – Lipophorin lipid could execute a crucial effect on Smoothened signaling. Thus, Patched is unlikely to be required for endosomal sterol efflux in general – it only influences the lipid composition of the subset of endosomes in which it is present and which constitute a special subcellular compartment different from the endocytic route utilized for the neutral lipid storage. In agreement with that, it has previously been shown that blocking the transport of bulk nutritional lipids by affecting the vesicular trafficking of NPC-1 does not significantly interfere with the Hedgehog signaling pathway (Incardona et al., 2000). We therefore hypothesize that Patched-mediated sequestration of Lipophorin diverts these particles away from trafficking pathways that promote neutral lipid storage. Consistent with that theory, Patched has recently been reported to reduce the accumulation of neutral lipid in a manner that does not depend on its SSD. Rather, this effect of Patched appears to depend on its ability to sequester Lipophorin (Callejo et al., 2008).

Interestingly, members of another family of MDR efflux pumps – ATP-binding cassette (ABC) transporters – have recently been shown to participate in reverse cholesterol transport and thereby modulate the assembly of the lipid rafts (Fitzgerald et al., 2010). In a similar manner, Patched might contribute to organization of a special lipid environment in the endosomes. Based on the analogy to the primary cilium in vertebrates, it is tempting to speculate that *Drosophila* cells might have developed a similar specialized compartment, which could serve as a recruitment platform for the components of the Hedgehog signaling pathway. The unique lipid composition of this compartment, generated by the action of the SSD of Patched, could alter Smoothened conformation, in a way that would favor its degradation. This idea correlates with the generally accepted theory that Smoothened and other members of the Hedgehog signaling pathway localize to microtubule-associated complexes scaffolded by Costal 2 (see also **Fig. 3.2**).

However, it is clear that other regions of Patched protein are equally important to provide efficient Smoothened destabilization. Thus, free addition Lipophorin lipids to the cells, which only express the mutant Patched^{SSD}, can circumvent the requirement for the SSD function, since the lipids are already mobilized in liposomes, and deplete Smoothened from the basolateral membrane (**Fig. 5.9** E-H and J). However, the same treatment is not sufficient to destabilize Smoothened in the total absence of Patched protein in *patched^{llw}* tissue (**Fig. 5.9** A-D and I). This result is not surprising and correlates with an earlier observation that accurate regulation of Smoothened trafficking by Patched requires an intact C-terminal tail – a region of Patched, which is responsible for its correct internalization and turnover (Lu et al., 2006). Further, we observed that Patched^{1130X} – a mutant form of Patched protein completely missing the C-terminal tail – was no longer able to stabilize Lipophorin in endosomes, showing that this region of Patched is required for correct Lipophorin sequestration (**Fig. 5.11** A-C). Notably, the ability of Patched to recruit Lipophorin is not mediated by Hedgehog, since Patched$^{\Delta loop2}$ is unable to bind Hedgehog, but nevertheless efficiently sequesters Lipophorin in endosomes (**Fig. 5.11** G-I). Based on these data, one could imagine that Patched cycles between the plasma membrane and a specific internal compartment, where it encounters internalized Lipophorin particles – a process dependent on the C-terminal region of Patched. Once sequestered, Lipophorin particles serve as a source of lipids needed for Smoothened destabilization, mobilized by the SSD of Patched. Thereby, the lipid composition of the endosome where Patched, Lipophorin and Smoothened localize, is altered, potentially inducing a change in Smoothened conformation and/or recruitment of cytoplasmic effectors facilitating its degradation – this process requires an intact C-terminal tail and possibly other regions of Patched protein.

6.5 Lipid candidates for Smoothened repression

Which Lipophorin-derived lipids does Patched mobilize to regulate Smoothened trafficking? We started to identify the active compound by fractionation on a TLC. Thereby, lipid species were separated based on their hydrophobicity and the abundant lipid bands were tested for their activity to destabilize Smoothened. Interestingly, our analysis identified one specific lipid fraction – extracted from the TLC band no. 3 – to have retained the full potential to reduce basolateral Smoothened levels when added to the explanted wing discs (**Fig. 5.19**). Preliminary results from the analysis of the lipid composition by mass spectroscopy suggest that ceramides and/or possibly other sphingolipids constitute the main lipid components of this fraction, as well as of the non-saponifiable lipid fraction, which also has a potential to reduce basolateral Smoothened levels (**Fig. 5.20**). Sphingolipids are attractive candidates, since many of them have been shown to participate in regulation of receptor signaling and play significant roles in various cellular processes, such as cell proliferation, apoptosis, cell cycle arrest, senescence, and inflammation (Ogretmen and Hannun, 2004; Ozbayraktar and Ulgen, 2009). However, commercially available ceramides did not show any activity in our assay so far (not shown). The reason for that might originate from structural differences of *Drosophila* lipids, which have several specific molecular characteristics, whereas most of the commercially available lipid species mainly resemble vertebrate lipids (Jones et al., 1992; Wiegandt, 1992; Rietveld et al., 1999; van Meer et al., 2008). Another possibility is that ceramides, being the most abundant lipid species in that fraction, might mask the presence of the actual active lipid. For instance, the region of the TLC where band no. 3 is located also corresponds to the mobility region of specifically modified, e.g. hydroxylated, sterols. Generally, sterols are attractive candidates for the role in Smoothened regulation, since they are present in Lipophorin particles and the SSD of Patched is able to regulate sterol trafficking from Patched endosomes. Nevertheless, our data do not support the possibility for bulk membrane sterol to affect Smoothened, since dietary sterol depletion neither alters Smoothened levels on the basolateral membrane, nor interferes with other aspects of Hedgehog signaling. Futhermore, addition of ergosterol, the most abundant membrane sterol in *Drosophila* (Rietveld et al., 1999), does not reduce Smoothened accumulation (**Fig. 5.7**). However, our data do not rule out the possibility that a specifically modified sterol species or a unique *Drosophila* sterol derivative might act at low concentrations to affect Smoothened signaling. Also, it is possible that several lipid species comprise a potential to inhibit Smoothened and might even act in conjunction. This idea is also consistent with the fact that the transporters of the RND family of permeases, which Patched and NPC-1 are members of, have rather broad substrate specificities (Piddock, 2006). For instance, it has been shown for NPC-1, that mutation of its SSD alters endosomal trafficking of many different lipids – sphingolipids, lysophospholipids and sterols being some examples of those (Wojtanik and Liscum, 2003; Ikonen and Holtta-Vuori, 2004; Mukherjee and Maxfield, 2004). Further fractionation as well as a more detailed analysis of the lipid compounds in the effective lipid fraction would shed light on the identity of the active lipid species.

6.6 Possibilities in Smoothened regulation

One relevant possibility for Smoothened regulation results from the fact that mammalian Smoothened can be affected by binding of different lipid species – agonists and antagonists (Frank-Kamenetsky et al., 2002; Chen et al., 2002a; Chen et al., 2002b;

Bijlsma et al., 2006). Thus, activation of Smoothened might be achieved by an action of an agonist, availability or presentation of which might be negatively regulated by the SSD of Patched. Indeed, Smoothened activity and its subcellular localization has been observed to be positively influenced by oxysterols and several synthetic compounds (Corcoran and Scott, 2006; Dwyer et al., 2007).

In fact, this hypothesis suggests a plausible mechanism for counteraction of Hedgehog with the repressive action of Patched on Smoothened (see also **Fig. 6.2**). Previously, Hedgehog has been shown to move throughout the wing disc tissue, being anchored in the phospholipid monolayer of Lipophorin particles (Panáková et al., 2005). Thus, a portion of internalized Lipophorin particles carries Hedgehog molecules on them, being mostly concentrated in the region near the Hedgehog-producing cell of the posterior compartment of the wing disc. Further away from the anteroposterior compartment boundary, free Lipophorin is more abundant than Hedgehog-transporting Lipophorin. There, Lipophorin is internalized independently of Hedgehog and, once reached the Patched-positive and Smoothened-containing endosomes, its contents can be utilized by Patched SSD and can act on Smoothened.

Close to the A/P boundary, where the concentration of Hedgehog on Lipophorin is the highest, the complex is internalized by binding of Hedgehog to Patched. This interaction increases Patched degradation (Incardona et al., 2000) and may thereby prevent Lipophorin sequestration and lipid mobilization. Alternatively, binding of Hedgehog could change the conformation of Patched in a way that would interfere with the function of Patched SSD. Thus, Hedgehog is thought to bind to the extracellular loops of Patched (Marigo et al., 1996). This is exactly the region, which is important for conferring substrate specificity in RND family transporters (Elkins and Nikaido, 2002; Mao et al., 2002). This facts suggest that Hedgehog binding to Patched might prevent the mobilization of Smoothened inhibitory lipid from the Lipophorin particles. At the same time, it might allow the access of Smoothened to a potential agonist. Thus, regulation of Smoothened might involve both – its Patched- and Lipophorin lipid-mediated degradation and/or its exhibition to an agonist, possibly present among the membrane lipids upon blocking the SSD activity of Patched by Hedgehog.

6.7 Further transduction of Smoothened signaling

In general, Smoothened activity – modulated by the lipid composition of the endosomal compartment – might subsequently regulate its signal transduction solely through recruitment of different effector components of the Hedgehog signaling pathway. This is a likely possibility, based on the fact that other members of the family of seven-transmembrane receptors act in a similar way to transmit the signal. Thus, these proteins employ their extracellular domain, sometimes in association with the transmembrane helices, for binding of a potential ligand, whereas the helical region functions as a conformational switch to regulate the recruitment of cytoplasmic effectors to the C-terminal cytoplasmic domain (**Fig. 6.1**). Thereby, different ligands can trigger activation of different signal transduction pathways (Perez and Karnik, 2005). Consequently, multiple effector proteins and the receptor itself assemble into a complex, which propagates the signal. Finally, internalization of the activated receptor leads to a desensitization of the signal (Pitcher et al., 1998).

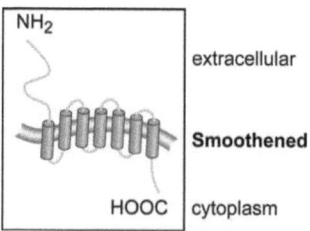

Figure 6.1 **Schematized structure of Smoothened.**

The N-terminal extracellular domain and the transmembrane helices (indicated in green) regions of Smo are highly conserved across phyla (adapted from (Ioannou, 2001)).

Interestingly, the N-terminal extracellular domain of Smoothened has been shown to be essential for the activity of the protein, since its deletion interferes with Smoothened signaling and causes mislocalization of the protein (Nakano et al., 2004; Aanstad et al., 2009). Furthermore, the helical region of the mammalian Smoothened has been observed to interact with agonist and antagonist molecules (Frank-Kamenetsky et al., 2002). Thus, further transduction of Hedgehog signaling may be regulated through recruitment of different effector proteins to Smoothened, which depends on its conformation influenced by an interaction with an activating ligand. Thereby, this ligand could be made available in Patched-positive endosomes when binding to Hedgehog blocks the repressive transporter activity of the Patched SSD.

6.8 Two possible modes of Smoothened signaling

Recently, mammalian Smoothened has been proposed to be activated in a two-step process. Thereby, Smoothened needs first to be translocated into the primary cilium, which, however, is insufficient for the induction of target gene transcription. This step can be repressed by cyclopamine. Full activation of Smoothened signaling is only achieved during the second step of induction by an unknown mechanism – a process controlled by Patched (Aanstad et al., 2009; Rohatgi et al., 2009).

Based on these arguments and our own observations described in this work, we hypothesize that Smoothened can actually achieve various states of signaling in *Drosophila* wing disc cells. Thus, its internalization into the endosomal compartment might favor its inactive state, being controlled by Patched and a Lipophorin-derived inhibitory lipid. Hedgehog-induced Patched degradation and/or its failure to sequester

Lipophorin and mobilize its lipids might partly activate Smoothened signaling, preventing its degradation. Further activation of Smoothened might be achieved by an interaction with an activating lipid compound, available in Patched-positive endosomes only in the presence of Hedgehog.

Consistent with this idea, loss of Lipophorin reproduces only a subset of the effects on Smoothened signaling – while it stabilizes Smoothened and causes Smoothened-dependent accumulation of full length Ci_{155}, it does not allow target gene activation (**Fig. 1**). Interestingly, a similar uncoupling of Ci_{155} stability and target gene activation has been already observed in *fused* and *dally* mutants (Alves et al., 1998; Wang and Holmgren, 1999; Eugster et al., 2007). Therefore, it is tempting to speculate that depletion of Lipophorin arrests Smoothened in a partly activated state.

How does this signaling mode of Smoothened promote Ci_{155} stabilization? In *Drosophila*, large portion of the latent full-length form Ci_{155} is anchored in the cytoplasm by the regulatory complex that is scaffolded by Costal 2 and includes the kinases PKA, CKI and GSK3β (Robbins et al., 1997; Sisson et al., 1997; Stegman et al., 2004; Zhang et al., 2005). When the kinase activity is abolished, Ci_{155} is not degraded. Interestingly, the phosphorylation sites of Ci resemble those primed by PKA, CKI and GSK3β in the C-terminal tail of Smoothened (Jia et al., 2004). This triggers the idea that partly activated Smoothened recruits these kinases to the endosomes and competes with Ci_{155} for a binding partner, thereby interfering with the cleavage and inducing stabilization of Ci_{155}.

However, full activation of Smoothened is necessary to promote further activation of Ci_{155} and induce the transcription of target genes. This step requires Hedgehog and cannot be accomplished in the wing disc cells of Lipophorin RNAi larvae, since Hedgehog is transported on Lipophorin and is therefore missing in the anterior compartment. The molecular basis of further Smoothened activation could simply be an alteration of its conformation. This change could either be induced by binding of an activating ligand or by phosphorylation and subsequent dimerization as it has been proposed earlier (Hooper, 2003). Thereby, modified lipid composition of the endosomes as a result of the alteration of Patched SSD action in the presence of Hedgehog might facilitate both – an interaction of Smoothened with such an agonist and/or phosphorylation of Smoothened, which would induce its full signaling activity as suggested previously (Zhao et al., 2007). Finally, the physical properties of the activator form of Ci_{155} have not been identified yet, but might involve a post-translational modification of the transcription factor (Ohlmeyer and Kalderon, 1998; Hooper and Scott, 2005). Future investigations, focusing on the subcellular localization and activity state of other members of Hedgehog signaling cascade, are therefore necessary to further elucidate these processes.

6.9 Applications

Since more than a decade now, the idea of a lipophilic Hedgehog pathway inhibitor, acting on Smoothened and catalytically regulated by Patched, is being explored. Resulting from many chemical screens, several natural and artificial compounds have been proven to have an inhibitory potential on Smoothened signaling. However, even though some derivatives of these molecules have now entered clinical trials as treatments for various forms of cancer (Rudin et al., 2009; Dierks, 2010), identification of an endogenous Smoothened inhibitor would be much more helpful in the development of anti-cancer drugs, regarding possible side effects. This work shows, for the first time, that one or more lipids derived from *Drosophila* Lipophorin act in conjunction with Patched to regulate Smoothened trafficking and activity *in vivo*.

Interestingly, preliminary results from our collaborative work with Philipp Beachy and colleagues indicate that Lipophorin lipids not only regulate Smoothened in *Drosophila*, but also have a similar repressive effect on the mammalian Smoothened.

Thus, Lipophorin lipid extract interferes with Smoothened signaling in *light2* cells (personal communication) – a cell culture model system to monitor mammalian Hedgehog signaling (Taipale et al., 2000). Moreover, Lipophorin lipids can apparently compete with cyclopamine for the binding to Smoothened in fibroblast cell culture (personal communication). These observations further emphasize the potential of Lipophorin lipids for anti-cancer drug development. Following fractionation of the Lipophorin lipid extract to identify the active lipid species is therefore crucial and would provide a solid basis for future implementations.

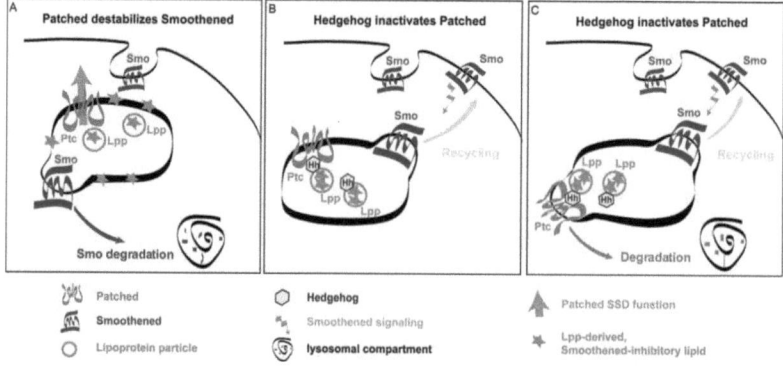

Figure 6.2 **A model for Patched-mediated Smoothened destabilization**
Smo (blue) is internalized and moves through endosomes that contain Ptc (red) and Lpp (pink). From these endosomes, Smo is either sorted to degradation or recycled to the basolateral membrane. This decision is controlled by Ptc. Ptc regulates the lipid composition of these endosomes via its SSD – partly by promoting the mobilization of lipids derived from Lpp (pink stars).
(A) Ptc destabilizes Smo. Here, Lpp lipids bias Smo trafficking towards degradation.
(B) Hh blocks lipid mobilization by Ptc. Here, Hh (yellow) on Lpp particles inhibits the mobilization of inhibitory lipids by Ptc. This elevates Smo levels on the basolateral membrane.
(C) Hh increases Ptc degradation. Here, Hh induces rapid degradation of Ptc and Lpp, thereby preventing Lpp sequestration and lipid mobilization. This elevates Smo levels on the basolateral membrane.

The results of this work enable us to suggest a working model, which is consistent with what is already known about the nature of Smoothened regulation by Patched. Thus, in the absence of Hedgehog, the transmembrane proteins Patched and Smoothened cycle between the basolateral membrane and the endosomes. At the same time, Lipophorin particles are internalized and either traffic along the degradative pathway that promotes neutral lipid storage or encounter the endosomes containing Patched. There, Lipophorin is stabilized, which allows Patched to mobilize Lipophorin-derived lipids via its SSD, thereby altering the lipid composition of these endosomes.

Smoothened that passes through Patched-containing endosomes can be targeted either for degradation or recycling, depending on the lipid composition of these endosomes. Thus, when Lipophorin-derived lipids, mobilized by Patched SSD, are available in this compartment, they bias Smoothened trafficking towards degradation. On the other hand, when the activity of Patched is impaired – either by dysfunctional SSD or through binding of Hedgehog to Patched – Lipophorin particles cannot be sequestered and Smoothened is targeted to the basolateral membrane. In summary, total Smoothened levels on the basolateral membrane and, consequently, its ability to stabilize Ci_{155} are regulated through a balance between Smoothened degradation versus its recycling, which is controlled by Patched and one or more Lipophorin-derived lipid species.

SUPPLEMENTS

7   SUPPLEMENTS

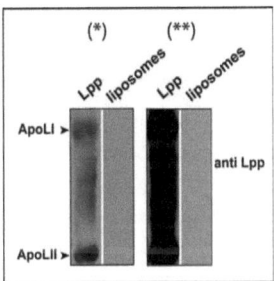

Figure 7.1   Protein-free derivation of Lipophorin lipids.
Western blot of a Lpp fraction and of an equivalent amount of liposomes constituted from Lpp lipids probed for both ApoLI and ApoLII. Both ApoLI and ApoLII are abundant in the Lpp fraction even at short exposure (*), but not in the lipid fraction, even at long exposure (**).

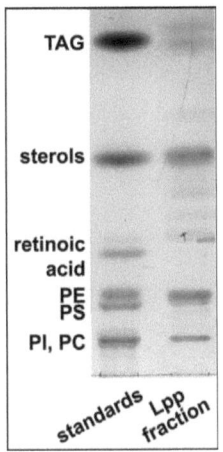

Figure 7.2   Major lipids present in Lipophorin.
Thin layer chromatograph of lipid extracts of Lpp particles with indicated standards. This experiment has been performed by Maria Joao Carvalho.

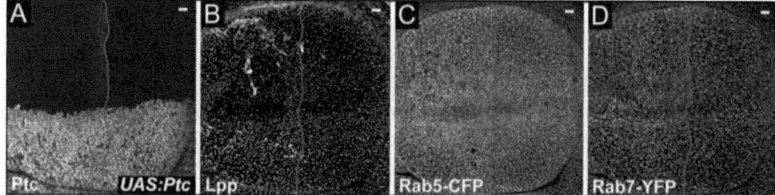

Figure 7.3 **Patched over-expression does not perturb endocytic compartments.**
(A-D) Apical region (0.7 - 2.8 µm below apical surface) of a wing disc over-expressing Ptc (A) in the dorsal compartment for 18 hours and ubiquitously expressing Rab5-CFP (C) and Rab7-YFP (D) has been stained for Lpp (B). Although Lpp is stabilized in Ptc over-expressing cells, neither Rab5 nor Rab7 positive endosomes are larger or more abundant in the dorsal compartment. Scale bar = 10 µm. A/P boundaries are indicated by blue lines.

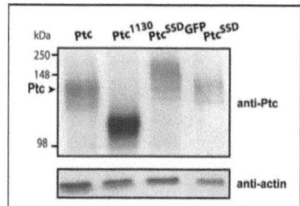

Figure 7.4 **Expression levels of different Patched alleles.**
Western blot of extracts of wing imaginal discs, which have been over-expressing Ptc, Ptc^{1130}, $Ptc^{SSD}GFP$ and Ptc^{SSD}. It has been probed for Patched and actin.

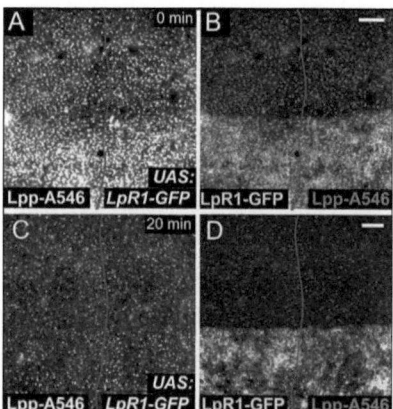

Figure 7.5 **LpR1 increases Lipophorin uptake, but does not affect Lipophorin degradation.**
(A-D) Apical region (0.7 - 2.8 µm below apical surface) of a wing disc over-expressing LpR1-GFP (A, C and B, D green) in the dorsal compartment has been incubated with Alexa546-labeled Lpp particles (B, D red) for 10 minutes and then either fixed immediately (A, B) or washed and subsequently incubated in Grace's medium for 20 minutes (C, D). The ratio of average staining intensity in the dorsal and ventral (Lpp_D / Lpp_V) compartments in pulse and chase is (Lpp_D / Lpp_V)$_{pulse}$ = 1.8 and (Lpp_D / Lpp_V)$_{chase}$ = 0.97, respectively. LpR1 facilitates Lpp uptake but does not decrease Lpp degradation.
Scale bars = 10 µm. A/P boundaries are indicated by blue lines.

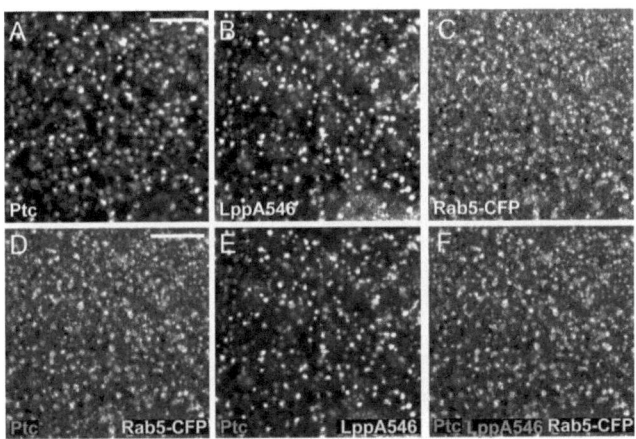

Figure 7.6 **Patched decreases Lipophorn degradation and stabilizes it in early endosomes.**
(A-F) Apical region (0.7 - 2.8 µm below apical surface) of a Ptc over-expressing wing disc ubiquitously expressing Rab5-CFP which was incubated with labeled Lpp particles for 10 minutes and fixed immediately. The image shows cells in the over-expressing compartment imaged for Ptc (A and D, E, F red), Lpp (B and E, green and F, blue) and Rab5-CFP (C and D, F green). Lpp and Ptc colocalize in Rab5-positive endosomes (in the dorsal compartment, 60.3 % of Ptc colocalizes with Lpp, 90.2 % of Ptc colocalizes with Rab5; 80.2 % of Lpp colocalizes with Ptc, 78.3 % of Lpp colocalizes with Rab5; 33.3 % of Rab5 colocalizes with Ptc, 21.0 % of Rab5 colocalizes with Lpp; (P-Values)$_{Costes}$ = 1.0).
Scale bars = 10 µm.

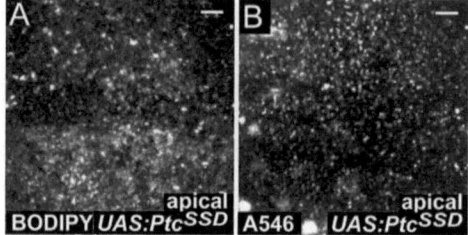

Figure 7.7 **Mutation of the Sterol-Sensing Domain of Patched perturbs derivation of lipid cargo from internalized Lipophorin particles.**

(A, B) apical region (0.7 - 2.8 μm below apical surface) of a wing disc from a larvae over-expressing Patched^{SSD} in the dorsal compartment, incubated for 40 minutes with purified Lipophorin particles double-labeled with BODIPY-cholesterol and Alexa546. Whereas Alexa546-labeled protein moiety is absent, BODIPY-cholesterol-labeled Lipophorin cargo accumulates in cells of the Patched^{SSD}-over-expressing compartment.

Scale bars = 10 μm.

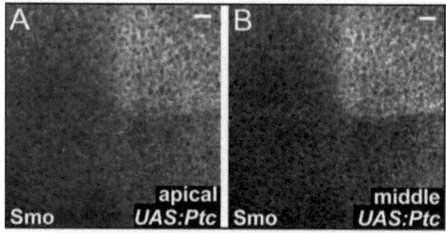

Figure 7.8 **Localization of Smoothened in Patched-over-expressing cells.**

(A) apical (0.7 - 2.8 μm below apical surface) and (B) middle (2.8 - 4.2 μm below apical surface) section of a wing disc over-expressing Ptc in the dorsal compartment, which has been stained for Smo. Ptc reduces both apical and basolateral Smo levels.

Scale bars = 10 μm.

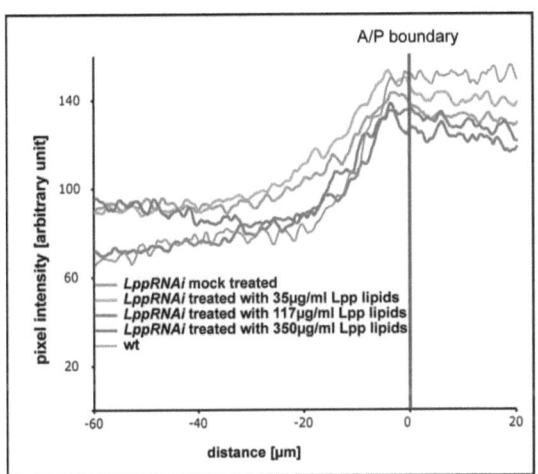

Figure 7.9 **Amount of Lipophorin-derived lipids necessary for reduction of basolateral Smoothened levels.**

Quantification of Smo staining of wing discs from LppRNAi larvae treated with Grace's medium or with liposomes prepared with 35 µg/ml, 117 µg/ml or 350 µg/ml of Lpp-derived lipids, compared to Smo staining intensity of wild type wing discs. Elevated Smo levels in LppRNAi wing discs are reversed to the wild type levels by treatment with 350 µg/ml Lpp-derived lipids (p<0.00006). At least 6 discs were quantified for each condition.

MATERIAL AND METHODS

MATERIAL AND METHODS

Drosophila stocks

The wild type Oregon R flies, as well as *hs*-flippase, *ap-GAL4*, *adh-GAL4*, *tub-GAL4* and *tubP-GAL80^{ts}* fly stocks are available from the Bloomington Stock Center.

Transgenic lines were obtained from: *UAS<HcRed>dsLpp* (Panáková et al., 2005), *UAS-Ptc* (Martin et al., 2001), *UAS-Ptc^{S2}GFP* (Torroja et al., 2004), *UASPtc^{SSD}* (Martin et al., 2001), *UAS-Ptc^{1130x}* (Johnson et al., 2000), *tubP-CFPRab5* (Marois et al., 2006), *UAS<HcRed>Rab7TN* (Marois et al., 2006).

Mutants: *ptc^{Iiw}* (also known as *Ptc^{16}*) (Strutt et al., 2001), *ptc^{S2}* (Martin et al., 2001).

Induction of RNAi

LppRNAi was induced as described (Panáková et al., 2005) using *Adh-GAL4* to drive fat body expression in larvae 48 hours after egg laying. Wing imaginal discs were dissected from larvae after 4 days of LppRNAi induction and analyzed further.

Clonal analysis

Ptc^{S2} and *ptc^{Iiw}* clones were generated as described (Martin et al., 2001).

Induction of Patched transgenes

Ap-GAL4, tubP-GAL80^{ts}/UAS-Ptc, *Ap-GAL4, tubP-GAL80^{ts}/UAS-Ptc^{SSD}*, *Ap-GAL4, tubP-GAL80^{ts}/UAS-Ptc^{S2}-GFP* and *Ap-GAL4, tubP-GAL80^{ts}/UAS-Ptc^{1130x}* animals were reared at 18°C, 3rd instar larvae were transferred to 29°C for 16 hours.

Antisera

Rabbit anti-Patched antibody was generated against the 2nd extracellular loop of the Patched protein and used at 1:300.

Immunohistochemistry

Imaginal discs were dissected in PBS at 4°C, fixed in 4 % paraformaldehyde for 20 minutes and permeabilized with 0.05 % Triton X-100 in PBS (PBT) twice for 10 minutes. The imaginal discs were then blocked three times for 15 minutes in PBT + 1 mg/ml BSA + 250 mM NaCl and incubated overnight with the primary antibody in PBT + 1 mg/ml BSA. Subsequently, the discs were washed twice for 20 minutes with PBT + 1 mg/ml BSA, twice for 20 minutes with the blocking solution PBT + 1 mg/ml BSA + 4 % normal goat serum and then incubated for at least 2 hours with the secondary antibody diluted 1:1000 in the blocking solution. The antibody was removed by washing three times for 15 minutes with PBT and three times for 15 minutes in PBS. Finally, the discs were mounted with Prolong Anti Fade reagent (Molecular Probes).

Primary antibodies were diluted as follows: rat anti-Ci 2A1 1:10 (Wang and Holmgren, 1999), mouse anti-Patched 1:100 (DSHB, University of Iowa, Department of Biological Sciences, Iowa City, IA 52242.), mouse anti-Smoothened 1:50 (DSHB, University of Iowa, Department of Biological Sciences, Iowa City, IA 52242.), rabbit anti-Hedgehog 1:500 and guinea pig anti-Lipophorin 1:1000 (Eugster et al., 2007).

Lipid depletion

The lipid depleted feeding medium was prepared with 10% chloroform-extracted yeast autolysate (Sigma), 10% glucose, 1% chloroform-extracted agarose and 0,015% Nipagen in water. The ingredients were mixed and boiled. Chloroform extraction of yeast autolysate and agarose was performed by incubation of the substances in 3 volumes of chloroform (Fluka) overnight and subsequent washing with 3 volumes of chloroform for four hours. Then, the substances were filtered through Whatman paper and air-dried. Early 2nd instar larvae were transferred from normal food to lipid-depleted medium for 4 days. Membrane sterol levels were visualized by filipin staining.

Feeding with BODIPY-cholesterol

Ap-GAL4, tubP-GAL80^{ts}/UAS-Ptc larvae were reared at 18°C until they reached the 2nd instar. Then they were transferred from normal food to the lipid-depleted medium containing 6.5 µg/ml BODIPY-cholesterol (Holtta-Vuori et al., 2008) and reared at 29°C for 16 hours. Then, the larvae were dissected in Grace's medium (Sigma) at room temperature and their imaginal discs were transferred onto glass slides into drops of medium delimited by a chamber of double-sided adhesive tape (Greco et al., 2001). The discs were oriented with the apical surface facing the cover slip and subjected to live imaging using Zeiss LSM 510 confocal microscope.

Filipin staining

Imaginal discs were dissected in PBS at 4°C, fixed in 4 % paraformaldehyde for 20 minutes and washes twice for 10 minutes in PBS. Then, the discs were stained with the 50 µg/ml Filipin solution (Sigma) for 30 minutes, washed twice for 10 minutes with PBS and subsequently mounted with Prolong Anti Fade reagent (Molecular Probes).

Image analysis

All quantified immunostaining was performed on discs that were dissected, fixed, stained and imaged in parallel using the same microscope settings. To quantify Ci and Smothened staining intensities, three apical sections 0.7 µm apart were projected using maximal intensity in ImageJ. For each image, two rectangles of 100 pixels parallel to the

A/P axis by 351 pixels parallel to the D/V axis were selected and centered at the A/P boundary in ventral and dorsal compartments. Average pixel intensity was determined as a function of distance from A/P boundary using PlotProfile and plotted using Microsoft Excel. All A/P boundaries were determined according to anti-Ci or anti-Patched co-immunostaining.

To estimate the significance of changes in staining intensities in discs of different genotypes, we measured Smoothened or Ci staining intensity at the same distance from the A/P boundary in each disc and calculated p values using Excel.

To quantify percent colocalization we used the colocalization plugin of the Fiji image processing software. Statistical significance was determined according to (Costes et al., 2004), where (P-Value)$_{Costes}$ = (1 – the fraction of randomized images giving a Pearson correlation coefficient greater than the original image).

Western blotting

Membranes were incubated with mouse anti-Patched (DSHB, University of Iowa, Department of Biological Sciences, Iowa City, IA 52242.) 1:100 and anti-actin (Sigma) 1:1000; followed by HRP conjugated anti-mouse IgM 1:5000 (Dianova) and anti-rabbit IgG 1:5000 (Dianova), respectively.

Isolation of Lipophorin particles

Lipophorin particles were isolated from wild type 3rd instar larvae by first homogenizing them in TNE buffer (100 mM Tris-Cl pH 7.5, 150 mM NaCl, 0.2 mM EGTA) plus protease inhibitors (Roche). The larvae were then broken with a loose pestle and the tissues and larval carcasses were pelleted by centrifugation at 1000g for 10 minutes. Supernatants were centrifuged at 33,600 rpm for 3 hours in a SW40Ti rotor (Beckman) to remove debris. Subsequently, KBr (Merck) was added to the sample to a concentration of 0.33

g/ml. The mixture was centrifuged at 39,000 rpm for 64 hours in a SW40Ti rotor. The top fraction (1.5 ml) of the density gradient, containing yellow-coloured Lipophorin particles, was removed and desalted on a Sephadex G-25 PD-10 column (Amersham Pharmacia Biotech) to replace TNE + KBr with PBS. Prior to that, the columns were equilibrated with 5 volumes of PBS.

Labeling of Lipophorin particles with Alexa546 succinimidyl ester probe

The labeling procedure was adapted from the Molecular Probes product information sheet for Amine-reactive probes.

1) preparation of Alexa546

1 mg of the Alexa546 dye (Molecular Probes) was dissolved in 100 μl of DMSO to a final concentration of 10 μg/μl. Aliquots of 10 μl were dried in the speed vacuum and immediately used or stored at -80 °C.

2) labeling

1 volume of desalted Lipophorin particles was mixed with 0.1 volume of 1M $NaHCO_3$, pH 7.3-9 in order to maintain the amine groups of Apolipophorin in a non-protonated state. An aliquot of Alexa546 dye was dissolved in 10 μl of DMSO (Sigma) and slowly added to the Lipophorin particles, while constantly vortexing the sample. The reaction was agitated at room temperature for at least one hour. Labeled Lipophorin particles were separated from the unreacted dye using Sephadex G-25 PD-10 columns and eluted with PBS. Prior to that, the columns were equilibrated with 5 volumes of PBS.

Uptake assay with labeled Lipophorin

Desalted Lipophorin particles were labeled with amine-reactive Alexa-546 probe as described above. Ap-GAL4, *tubP-GAL80^{ts}*/UAS-Ptc larvae, reared at 29°C for 18 hours, were dissected in Grace's medium (Sigma) and incubated with labeled Lipophorin particles (at a protein concentration of 0.5 mg/ml in Grace's medium) for 10 minutes at 22°C. 10 minutes is the minimum time required to see uptake of labeled Dextran by imaginal disc cells (Marois et al., 2006). The discs were then either washed immediately in PBS at 4°C and fixed (4% PFA for 20 minutes at room temperature), or washed and incubated for varying times in Grace's medium at 22°C before fixation.

Two-step Bligh and Dyer method for lipid extraction

Lipids were extracted from purified Lipophorin particles by a two-step Bligh and Dyer method (Bligh and Dyer, 1959). Thereby, 1 volume of purified Lipophorin particles was agitated for one hour with 3.75 volumes of chloroform (Fluka) : methanol (Merck), mixed in a proportion of 1:2. Then, 1.25 volume of chloroform was added and vortexed for one minute. Then, 1.25 volume of water was added and vortexed for one minute. The sample was centrifuged for 20 minutes at 1000 rpm and the lower (organic) phase was collected to a clean glass tube. The upper phase was mixed with 1.88 volumes of chloroform, vortexed and centrifuged for 20 minutes at 1000 rpm. The lower phase was collected and combined with the previously collected one. The solvent was evaporated under a flow of nitrogen and the lipid sample either subjected to immediate analysis or stored at -20°C.

Lipid quantification by charring

Total lipid concentration was measured by a method modified from (Marsh and Weinstein, 1966). Cholesterol solutions of 15, 30, 60, 90 and 120 µg were used as lipid standards. The standards and the lipid samples were placed in clean glass tubes and the solvent was evaporated from the samples by heating them at 120°C. Then, 2 ml of sulfuric acid was added to each sample and incubated for 15 minutes at 200°C. The samples were cooled by placing the tubes for 15 seconds in water at room temperature and then transferred to an ice bath for 5 minutes. Subsequently, 3 ml of water was added to each sample, the contents were mixed thoroughly and the samples were replaced on ice. When cool, the optical density was photometrically measured at the wavelength of 375 nm.

Uptake assay with liposomes

Liposomes were prepared from dried lipids by sonication for 30 minutes into Grace's medium to produce a final total lipid concentration of 500 µg/ml. Assuming that hemolymph represents approximately half of the larval volume, this concentration should be similar to that of lipids contributed by the Lipophorin fraction of the hemolymph. Wing imaginal discs from *hsp70-flp/+; Adh-GAL4/+; UAS<HcRed>dsLpp/ +* larvae were incubated in Grace's medium or in Grace's medium + liposomes for 2 hours at 22°C. Discs were fixed and immunostained as previously described.

Uptake assay with liposomes and Red Dextran

3rd instar larvae were dissected at 25°C in Grace's insect medium. To stain the endocytic compartments, imaginal discs from *hsp70-flp/+; Adh-GAL4/+; UAS<HcRed>dsLpp/ +* larvae were incubated in 10,000 MW lysine fixable Red Dextran diluted 1:10 in Grace's medium or in 10,000 MW lysine fixable Red Dextran diluted 1:10 in Grace's medium + liposomes for 2 hours at 22°C, then fixed and immunostained as previously described.

Saponification

To remove saponifiable lipids, dried Lipophorin lipid extract containing 500 nmol of total lipid was incubated at 80°C with 2 ml of 0.3 M methanolic potassium hydroxide for one hour. After cooling, the non-saponifiable fraction was extracted by two washes with diethyl ether, the fractions were combined and run through weak anion exchange DEAE Sephadex A-50 column to remove the contaminating fatty acids.

Preparative Thin Layer Chromatography

Lipophorin lipid extract and relevant lipid standards were loaded on thin layer chromatography (TLC) silica plates (Merck) and run in two sequential solvent systems. Thereby, the plate was first run in a mixture of chloroform : triethylamine : ethanol : water (35 : 35 : 40 : 9), to separate the more hydrophilic lipids. Then, the plate was placed in a solvent mixture of isohexane : ethyl acetate (5 : 1) to achieve better separation of hydrophobic lipids according to (Kuerschner et al., 2005). For further analysis, the TLC plate was split in two halves – one contained the lipid standards and a stripe of the Lipophorin lipid extract sample and was detected by spraying with 20 % sulfuric acid and heating to 150-200°C for 10 minutes. The second TLC part contained the residual Lipophorin lipid sample and was used for the scraping and extraction of lipid bands, without performing the detection step.

Extraction of TLC bands

Lipophorin lipid bands of interest were scraped from the TLC plate into separate glass vials. Then, lipid was extracted from the silica by agitation with 500 µl of chloroform : methanol (2 : 1) for 20 minutes and the solvent was collected to a new tube. The pellet was re-extracted by agitation with 500 µl chloroform : methanol (2 : 1) and supernatant was combined with the previous one. To remove the silica, the sample was centrifuged at 14000 g for 15 minutes and supernatant was collected to a clean glass vial. The lipids were dried and immediately used for further experiments or stored at -80°C.

BIBLIOGRAPHY

9 BIBLIOGRAPHY

Aanstad, P., Santos, N., Corbit, K. C., Scherz, P. J., Trinh le, A., Salvenmoser, W., Huisken, J., Reiter, J. F. and Stainier, D. Y. (2009). The extracellular domain of Smoothened regulates ciliary localization and is required for high-level Hh signaling. *Curr Biol* **19**, 1034-9.

Alcedo, J., Ayzenzon, M., Von Ohlen, T., Noll, M. and Hooper, J. E. (1996). The Drosophila smoothened gene encodes a seven-pass membrane protein, a putative receptor for the hedgehog signal. *Cell* **86**, 221-32.

Alcedo, J. and Noll, M. (1997). Hedgehog and its patched-smoothened receptor complex: a novel signalling mechanism at the cell surface. *Biol Chem* **378**, 583-90.

Alexandre, C., Jacinto, A. and Ingham, P. W. (1996). Transcriptional activation of hedgehog target genes in Drosophila is mediated directly by the cubitus interruptus protein, a member of the GLI family of zinc finger DNA-binding proteins. *Genes Dev* **10**, 2003-13.

Alves, G., Limbourg-Bouchon, B., Tricoire, H., Brissard-Zahraoui, J., Lamour-Isnard, C. and Busson, D. (1998). Modulation of Hedgehog target gene expression by the Fused serine-threonine kinase in wing imaginal discs. *Mech Dev* **78**, 17-31.

Apionishev, S., Katanayeva, N. M., Marks, S. A., Kalderon, D. and Tomlinson, A. (2005). Drosophila Smoothened phosphorylation sites essential for Hedgehog signal transduction. *Nat Cell Biol* **7**, 86-92.

Arrese, E. L., Canavoso, L. E., Jouni, Z. E., Pennington, J. E., Tsuchida, K. and Wells, M. A. (2001). Lipid storage and mobilization in insects: current status and future directions. *Insect Biochem Mol Biol* **31**, 7-17.

Aza-Blanc, P. and Kornberg, T. B. (1999). Ci: a complex transducer of the hedgehog signal. *Trends Genet* **15**, 458-62.

Baron, M. H. (2003). Embryonic origins of mammalian hematopoiesis. *Exp Hematol* **31**, 1160-9.

Beenakkers, A. M., Van der Horst, D. J. and Van Marrewijk, W. J. (1985). Insect lipids and lipoproteins, and their role in physiological processes. *Prog Lipid Res* **24**, 19-67.

Bijlsma, M. F., Spek, C. A., Zivkovic, D., van de Water, S., Rezaee, F. and Peppelenbosch, M. P. (2006). Repression of smoothened by patched-dependent (pro-)vitamin D3 secretion. *PLoS Biol* **4**, e232.

Bischof, M. G., Heinze, G. and Vierhapper, H. (2006). Vitamin D status and its relation to age and body mass index. *Horm Res* **66**, 211-5.

Bligh, E. G. and Dyer, W. J. (1959). A rapid method of total lipid extraction and purification. *Can. J. Biochem. Physiol.* **37**, 911-917.

Butler, M. J., Jacobsen, T. L., Cain, D. M., Jarman, M. G., Hubank, M., Whittle, J. R., Phillips, R. and Simcox, A. (2003). Discovery of genes with highly restricted expression patterns in the Drosophila wing disc using DNA oligonucleotide microarrays. *Development* **130**, 659-70.

Callejo, A., Culi, J. and Guerrero, I. (2008). Patched, the receptor of Hedgehog, is a lipoprotein receptor. *Proc Natl Acad Sci U S A* **105**, 912-7.

Callejo, A., Torroja, C., Quijada, L. and Guerrero, I. (2006). Hedgehog lipid modifications are required for Hedgehog stabilization in the extracellular matrix. *Development* **133**, 471-83.

Chang, D. T., Lopez, A., von Kessler, D. P., Chiang, C., Simandl, B. K., Zhao, R., Seldin, M. F., Fallon, J. F. and Beachy, P. A. (1994). Products, genetic linkage and limb patterning activity of a murine hedgehog gene. *Development* **120**, 3339-53.

Chavrier, P., Gorvel, J. P., Stelzer, E., Simons, K., Gruenberg, J. and Zerial, M. (1991). Hypervariable C-terminal domain of rab proteins acts as a targeting signal. *Nature* **353**, 769-72.

Chen, J. K., Taipale, J., Cooper, M. K. and Beachy, P. A. (2002a). Inhibition of Hedgehog signaling by direct binding of cyclopamine to Smoothened. *Genes Dev* **16**, 2743-8.

Chen, J. K., Taipale, J., Young, K. E., Maiti, T. and Beachy, P. A. (2002b). Small molecule modulation of Smoothened activity. *Proc Natl Acad Sci U S A* **99**, 14071-6.

Chen, Y. and Struhl, G. (1996). Dual roles for patched in sequestering and transducing Hedgehog. *Cell* **87**, 553-63.

Chen, Y. and Struhl, G. (1998). In vivo evidence that Patched and Smoothened constitute distinct binding and transducing components of a Hedgehog receptor complex. *Development* **125**, 4943-8.

Choudhury, A., Sharma, D. K., Marks, D. L. and Pagano, R. E. (2004). Elevated endosomal cholesterol levels in Niemann-Pick cells inhibit rab4 and perturb membrane recycling. *Mol Biol Cell* **15**, 4500-11.

Clayton, R. B. (1964). The Utilization of Sterols by Insects. *J Lipid Res* **15**, 3-19.

Clement, V., Sanchez, P., de Tribolet, N., Radovanovic, I. and Ruiz i Altaba, A. (2007). HEDGEHOG-GLI1 signaling regulates human glioma growth, cancer stem cell self-renewal, and tumorigenicity. *Curr Biol* **17**, 165-72.

Corbit, K. C., Aanstad, P., Singla, V., Norman, A. R., Stainier, D. Y. and Reiter, J. F. (2005). Vertebrate Smoothened functions at the primary cilium. *Nature* **437**, 1018-21.

Corcoran, R. B. and Scott, M. P. (2006). Oxysterols stimulate Sonic hedgehog signal transduction and proliferation of medulloblastoma cells. *Proc Natl Acad Sci U S A* **103**, 8408-13.

Costes, S. V., Daelemans, D., Cho, E. H., Dobbin, Z., Pavlakis, G. and Lockett, S. (2004). Automatic and quantitative measurement of protein-protein colocalization in live cells. *Biophys J* **86**, 3993-4003.

Dellovade, T., Romer, J. T., Curran, T. and Rubin, L. L. (2006). The hedgehog pathway and neurological disorders. *Annu Rev Neurosci* **29**, 539-63.

DeLuca, H. F. (2004). Overview of general physiologic features and functions of vitamin D. *Am J Clin Nutr* **80**, 1689S-96S.

Denef, N., Neubuser, D., Perez, L. and Cohen, S. M. (2000). Hedgehog induces opposite changes in turnover and subcellular localization of patched and smoothened. *Cell* **102**, 521-31.

Dierks, C. (2010). GDC-0449--targeting the hedgehog signaling pathway. *Recent Results Cancer Res* **184**, 235-8.

Dwyer, J. R., Sever, N., Carlson, M., Nelson, S. F., Beachy, P. A. and Parhami, F. (2007). Oxysterols are novel activators of the hedgehog signaling pathway in pluripotent mesenchymal cells. *J Biol Chem* **282**, 8959-68.

Eaton, S. (2006). Release and trafficking of lipid-linked morphogens. *Curr Opin Genet Dev* **16**, 17-22.

Echelard, Y., Epstein, D. J., St-Jacques, B., Shen, L., Mohler, J., McMahon, J. A. and McMahon, A. P. (1993). Sonic hedgehog, a member of a family of putative signaling molecules, is implicated in the regulation of CNS polarity. *Cell* **75**, 1417-30.

Elkins, C. A. and Nikaido, H. (2002). Substrate specificity of the RND-type multidrug efflux pumps AcrB and AcrD of Escherichia coli is determined predominantly by two large periplasmic loops. *J Bacteriol* **184**, 6490-8.

Eswaran, J., Koronakis, E., Higgins, M. K., Hughes, C. and Koronakis, V. (2004). Three's company: component structures bring a closer view of tripartite drug efflux pumps. *Curr Opin Struct Biol* **14**, 741-7.

Eugster, C., Panakova, D., Mahmoud, A. and Eaton, S. (2007). Lipoprotein-heparan sulfate interactions in the Hh pathway. *Dev Cell* **13**, 57-71.

Fitzgerald, M. L., Mujawar, Z. and Tamehiro, N. (2010). ABC transporters, atherosclerosis and inflammation. *Atherosclerosis*.

Frank-Kamenetsky, M., Zhang, X. M., Bottega, S., Guicherit, O., Wichterle, H., Dudek, H., Bumcrot, D., Wang, F. Y., Jones, S., Shulok, J. et al. (2002). Small-molecule modulators of Hedgehog signaling: identification and characterization of Smoothened agonists and antagonists. *J Biol* **1**, 10.

Ganley, I. G. and Pfeffer, S. R. (2006). Cholesterol accumulation sequesters Rab9 and disrupts late endosome function in NPC1-deficient cells. *J Biol Chem* **281**, 17890-9.

Gerdes, J. M., Davis, E. E. and Katsanis, N. (2009). The vertebrate primary cilium in development, homeostasis, and disease. *Cell* **137**, 32-45.

Gimpl, G., Burger, K. and Fahrenholz, F. (1997). Cholesterol as modulator of receptor function. *Biochemistry* **36**, 10959-74.

Goodrich, L. V., Johnson, R. L., Milenkovic, L., McMahon, J. A. and Scott, M. P. (1996). Conservation of the hedgehog/patched signaling pathway from flies to mice: induction of a mouse patched gene by Hedgehog. *Genes Dev* **10**, 301-12.

Greco, V., Hannus, M. and Eaton, S. (2001). Argosomes: a potential vehicle for the spread of morphogens through epithelia. *Cell* **106**, 633-45.

Guerrero, I. and Chiang, C. (2007). A conserved mechanism of Hedgehog gradient formation by lipid modifications. *Trends Cell Biol* **17**, 1-5.

Haycraft, C. J., Banizs, B., Aydin-Son, Y., Zhang, Q., Michaud, E. J. and Yoder, B. K. (2005). Gli2 and Gli3 localize to cilia and require the intraflagellar transport protein polaris for processing and function. *PLoS Genet* **1**, e53.

Holtta-Vuori, M., Uronen, R. L., Repakova, J., Salonen, E., Vattulainen, I., Panula, P., Li, Z., Bittman, R. and Ikonen, E. (2008). BODIPY-cholesterol: a new tool to visualize sterol trafficking in living cells and organisms. *Traffic* **9**, 1839-49.

Hooper, J. E. (2003). Smoothened translates Hedgehog levels into distinct responses. *Development* **130**, 3951-63.

Hooper, J. E. and Scott, M. P. (2005). Communicating with Hedgehogs. *Nat Rev Mol Cell Biol* **6**, 306-17.

Huangfu, D. and Anderson, K. V. (2005). Cilia and Hedgehog responsiveness in the mouse. *Proc Natl Acad Sci U S A* **102**, 11325-30.

Huangfu, D. and Anderson, K. V. (2006). Signaling from Smo to Ci/Gli: conservation and divergence of Hedgehog pathways from Drosophila to vertebrates. *Development* **133**, 3-14.

Ikonen, E. and Holtta-Vuori, M. (2004). Cellular pathology of Niemann-Pick type C disease. *Semin Cell Dev Biol* **15**, 445-54.

Incardona, J. P., Gaffield, W., Lange, Y., Cooney, A., Pentchev, P. G., Liu, S., Watson, J. A., Kapur, R. P. and Roelink, H. (2000). Cyclopamine inhibition of Sonic hedgehog signal transduction is not mediated through effects on cholesterol transport. *Dev Biol* **224**, 440-52.

Incardona, J. P., Gruenberg, J. and Roelink, H. (2002). Sonic hedgehog induces the segregation of patched and smoothened in endosomes. *Curr Biol* **12**, 983-95.

Infante, R. E., Abi-Mosleh, L., Radhakrishnan, A., Dale, J. D., Brown, M. S. and Goldstein, J. L. (2008a). Purified NPC1 protein. I. Binding of cholesterol and oxysterols to a 1278-amino acid membrane protein. *J Biol Chem* **283**, 1052-63.

Infante, R. E., Radhakrishnan, A., Abi-Mosleh, L., Kinch, L. N., Wang, M. L., Grishin, N. V., Goldstein, J. L. and Brown, M. S. (2008b). Purified NPC1 protein: II. Localization of sterol binding to a 240-amino acid soluble luminal loop. *J Biol Chem* **283**, 1064-75.

Infante, R. E., Wang, M. L., Radhakrishnan, A., Kwon, H. J., Brown, M. S. and Goldstein, J. L. (2008c). NPC2 facilitates bidirectional transfer of cholesterol between NPC1 and lipid bilayers, a step in cholesterol egress from lysosomes. *Proc Natl Acad Sci U S A* **105**, 15287-92.

Ingham, P. W. and Placzek, M. (2006). Orchestrating ontogenesis: variations on a theme by sonic hedgehog. *Nat Rev Genet* **7**, 841-50.

Ingham, P. W., Taylor, A. M. and Nakano, Y. (1991). Role of the Drosophila patched gene in positional signalling. *Nature* **353**, 184-7.

Ioannou, Y. A. (2001). Multidrug permeases and subcellular cholesterol transport. *Nat Rev Mol Cell Biol* **2**, 657-68.

Ishizuya-Oka, A. and Hasebe, T. (2008). Sonic hedgehog and bone morphogenetic protein-4 signaling pathway involved in epithelial cell renewal along the radial axis of the intestine. *Digestion* **77 Suppl 1**, 42-7.

Jia, J., Tong, C. and Jiang, J. (2003). Smoothened transduces Hedgehog signal by physically interacting with Costal2/Fused complex through its C-terminal tail. *Genes Dev* **17**, 2709-20.

Jia, J., Tong, C., Wang, B., Luo, L. and Jiang, J. (2004). Hedgehog signalling activity of Smoothened requires phosphorylation by protein kinase A and casein kinase I. *Nature* **432**, 1045-50.

Johnson, R. L., Milenkovic, L. and Scott, M. P. (2000). In vivo functions of the patched protein: requirement of the C terminus for target gene inactivation but not Hedgehog sequestration. *Mol Cell* **6**, 467-78.

Johnson, R. L., Zhou, L. and Bailey, E. C. (2002). Distinct consequences of sterol sensor mutations in Drosophila and mouse patched homologs. *Dev Biol* **242**, 224-35.

Jones, H. E., Harwood, J. L., Bowen, I. D. and Griffiths, G. (1992). Lipid composition of subcellular membranes from larvae and prepupae of Drosophila melanogaster. *Lipids* **27**, 984-7.

Kalderon, D. (2005). The mechanism of hedgehog signal transduction. *Biochem Soc Trans* **33**, 1509-12.

Keeler, R. F. (1978). Cyclopamine and related steroidal alkaloid teratogens: their occurrence, structural relationship, and biologic effects. *Lipids* **13**, 708-15.

Keeler, R. F. and Balls, L. D. (1978). Teratogenic effects in cattle of Conium maculatum and conium alkaloids and analogs. *Clin Toxicol* **12**, 49-64.

Kent, D., Bush, E. W. and Hooper, J. E. (2006). Roadkill attenuates Hedgehog responses through degradation of Cubitus interruptus. *Development* **133**, 2001-10.

Khaliullina, H., Panakova, D., Eugster, C., Riedel, F., Carvalho, M. and Eaton, S. (2009). Patched regulates Smoothened trafficking using lipoprotein-derived lipids. *Development* **136**, 4111-21.

Krauss, S., Concordet, J. P. and Ingham, P. W. (1993). A functionally conserved homolog of the Drosophila segment polarity gene hh is expressed in tissues with polarizing activity in zebrafish embryos. *Cell* **75**, 1431-44.

Kuerschner, L., Ejsing, C. S., Ekroos, K., Shevchenko, A., Anderson, K. I. and Thiele, C. (2005). Polyene-lipids: a new tool to image lipids. *Nat Methods* **2**, 39-45.

Kutty, R. K., Kutty, G., Kambadur, R., Duncan, T., Koonin, E. V., Rodriguez, I. R., Odenwald, W. F. and Wiggert, B. (1996). Molecular characterization and developmental expression of a retinoid- and fatty acid-binding glycoprotein from Drosophila. A putative lipophorin. *J Biol Chem* **271**, 20641-9.

Kuwabara, P. E. and Labouesse, M. (2002). The sterol-sensing domain: multiple families, a unique role? *Trends Genet* **18**, 193-201.

Kwon, H. J., Abi-Mosleh, L., Wang, M. L., Deisenhofer, J., Goldstein, J. L., Brown, M. S. and Infante, R. E. (2009). Structure of N-terminal domain of NPC1 reveals distinct subdomains for binding and transfer of cholesterol. *Cell* **137**, 1213-24.

Lebrand, C., Corti, M., Goodson, H., Cosson, P., Cavalli, V., Mayran, N., Faure, J. and Gruenberg, J. (2002). Late endosome motility depends on lipids via the small GTPase Rab7. *EMBO J* **21**, 1289-300.

Lefers, M. A., Wang, Q. T. and Holmgren, R. A. (2001). Genetic dissection of the Drosophila Cubitus interruptus signaling complex. *Dev Biol* **236**, 411-20.

Lu, X., Liu, S. and Kornberg, T. B. (2006). The C-terminal tail of the Hedgehog receptor Patched regulates both localization and turnover. *Genes Dev* **20**, 2539-51.

Lum, L. and Beachy, P. A. (2004). The Hedgehog response network: sensors, switches, and routers. *Science* **304**, 1755-9.

Lum, L., Zhang, C., Oh, S., Mann, R. K., von Kessler, D. P., Taipale, J., Weis-Garcia, F., Gong, R., Wang, B. and Beachy, P. A. (2003). Hedgehog signal transduction via Smoothened association with a cytoplasmic complex scaffolded by the atypical kinesin, Costal-2. *Mol Cell* **12**, 1261-74.

Machold, R., Hayashi, S., Rutlin, M., Muzumdar, M. D., Nery, S., Corbin, J. G., Gritli-Linde, A., Dellovade, T., Porter, J. A., Rubin, L. L. et al. (2003). Sonic hedgehog is required for progenitor cell maintenance in telencephalic stem cell niches. *Neuron* **39**, 937-50.

Mann, R. K. and Beachy, P. A. (2004). Novel lipid modifications of secreted protein signals. *Annu Rev Biochem* **73**, 891-923.

Mao, W., Warren, M. S., Black, D. S., Satou, T., Murata, T., Nishino, T., Gotoh, N. and Lomovskaya, O. (2002). On the mechanism of substrate specificity by resistance nodulation division (RND)-type multidrug resistance pumps: the large periplasmic loops of MexD from Pseudomonas aeruginosa are involved in substrate recognition. *Mol Microbiol* **46**, 889-901.

Marigo, V., Davey, R. A., Zuo, Y., Cunningham, J. M. and Tabin, C. J. (1996). Biochemical evidence that patched is the Hedgehog receptor. *Nature* **384**, 176-9.

Marois, E., Mahmoud, A. and Eaton, S. (2006). The endocytic pathway and formation of the Wingless morphogen gradient. *Development* **133**, 307-17.

Marsh, J. B. and Weinstein, D. B. (1966). Simple charring method for determination of lipids. *J Lipid Res* **7**, 574-6.

Martin, V., Carrillo, G., Torroja, C. and Guerrero, I. (2001). The sterol-sensing domain of Patched protein seems to control Smoothened activity through Patched vesicular trafficking. *Curr Biol* **11**, 601-7.

Martinez Arias, A. (2003). Wnts as morphogens? The view from the wing of Drosophila. *Nat Rev Mol Cell Biol* **4**, 321-5.

McMahon, A. P., Ingham, P. W. and Tabin, C. J. (2003). Developmental roles and clinical significance of hedgehog signaling. *Curr Top Dev Biol* **53**, 1-114.

Merchant, M., Vajdos, F. F., Ultsch, M., Maun, H. R., Wendt, U., Cannon, J., Desmarais, W., Lazarus, R. A., de Vos, A. M. and de Sauvage, F. J. (2004). Suppressor of fused regulates Gli activity through a dual binding mechanism. *Mol Cell Biol* **24**, 8627-41.

Methot, N. and Basler, K. (2000). Suppressor of fused opposes hedgehog signal transduction by impeding nuclear accumulation of the activator form of Cubitus interruptus. *Development* **127**, 4001-10.

Milenkovic, L., Scott, M. P. and Rohatgi, R. (2009). Lateral transport of Smoothened from the plasma membrane to the membrane of the cilium. *J Cell Biol* **187**, 365-74.

Mukherjee, S. and Maxfield, F. R. (2004). Lipid and cholesterol trafficking in NPC. *Biochim Biophys Acta* **1685**, 28-37.

Murone, M., Rosenthal, A. and de Sauvage, F. J. (1999a). Hedgehog signal transduction: from flies to vertebrates. *Exp Cell Res* **253**, 25-33.

Murone, M., Rosenthal, A. and de Sauvage, F. J. (1999b). Sonic hedgehog signaling by the patched-smoothened receptor complex. *Curr Biol* **9**, 76-84.

Nakano, Y., Nystedt, S., Shivdasani, A. A., Strutt, H., Thomas, C. and Ingham, P. W. (2004). Functional domains and sub-cellular distribution of the Hedgehog transducing protein Smoothened in Drosophila. *Mech Dev* **121**, 507-18.

Neumann, S., Harterink, M. and Sprong, H. (2007). Hitch-hiking between cells on lipoprotein particles. *Traffic* **8**, 331-8.

Nybakken, K. and Perrimon, N. (2002a). Hedgehog signal transduction: recent findings. *Curr Opin Genet Dev* **12**, 503-11.

Nybakken, K. and Perrimon, N. (2002b). Heparan sulfate proteoglycan modulation of developmental signaling in Drosophila. *Biochim Biophys Acta* **1573**, 280-91.

Ogretmen, B. and Hannun, Y. A. (2004). Biologically active sphingolipids in cancer pathogenesis and treatment. *Nat Rev Cancer* **4**, 604-16.

Ohgami, N., Ko, D. C., Thomas, M., Scott, M. P., Chang, C. C. and Chang, T. Y. (2004). Binding between the Niemann-Pick C1 protein and a photoactivatable cholesterol analog requires a functional sterol-sensing domain. *Proc Natl Acad Sci U S A* **101**, 12473-8.

Ohlmeyer, J. T. and Kalderon, D. (1998). Hedgehog stimulates maturation of Cubitus interruptus into a labile transcriptional activator. *Nature* **396**, 749-53.

Ozbayraktar, F. B. and Ulgen, K. O. (2009). Molecular facets of sphingolipids: mediators of diseases. *Biotechnol J* **4**, 1028-41.

Palma, V., Lim, D. A., Dahmane, N., Sanchez, P., Brionne, T. C., Herzberg, C. D., Gitton, Y., Carleton, A., Alvarez-Buylla, A. and Ruiz i Altaba, A. (2005). Sonic hedgehog controls stem cell behavior in the postnatal and adult brain. *Development* **132**, 335-44.

Panáková, D., Sprong, H., Marois, E., Thiele, C. and Eaton, S. (2005). Lipoprotein particles are required for Hedgehog and Wingless signalling. *Nature* **435**, 58-65.

Pasca di Magliano, M. and Hebrok, M. (2003). Hedgehog signalling in cancer formation and maintenance. *Nat Rev Cancer* **3**, 903-11.

Pepinsky, R. B., Zeng, C., Wen, D., Rayhorn, P., Baker, D. P., Williams, K. P., Bixler, S. A., Ambrose, C. M., Garber, E. A., Miatkowski, K. et al. (1998). Identification of a palmitic acid-modified form of human Sonic hedgehog. *J Biol Chem* **273**, 14037-45.

Perez, D. M. and Karnik, S. S. (2005). Multiple signaling states of G-protein-coupled receptors. *Pharmacol Rev* **57**, 147-61.

Peters, C., Wolf, A., Wagner, M., Kuhlmann, J. and Waldmann, H. (2004). The cholesterol membrane anchor of the Hedgehog protein confers stable membrane association to lipid-modified proteins. *Proc Natl Acad Sci U S A* **101**, 8531-6.

Philipp, M., Fralish, G. B., Meloni, A. R., Chen, W., MacInnes, A. W., Barak, L. S. and Caron, M. G. (2008). Smoothened signaling in vertebrates is facilitated by a G protein-coupled receptor kinase. *Mol Biol Cell* **19**, 5478-89.

Pho, D. B., Pennanec'h, M. and Jallon, J. M. (1996). Purification of adult Drosophila melanogaster lipophorin and its role in hydrocarbon transport. *Arch Insect Biochem Physiol* **31**, 289-303.

Piddock, L. J. (2006). Multidrug-resistance efflux pumps - not just for resistance. *Nat Rev Microbiol* **4**, 629-36.

Pitcher, J. A., Freedman, N. J. and Lefkowitz, R. J. (1998). G protein-coupled receptor kinases. *Annu Rev Biochem* **67**, 653-92.

Porter, J. A., Ekker, S. C., Park, W. J., von Kessler, D. P., Young, K. E., Chen, C. H., Ma, Y., Woods, A. S., Cotter, R. J., Koonin, E. V. et al. (1996a). Hedgehog patterning activity: role of a lipophilic modification mediated by the carboxy-terminal autoprocessing domain. *Cell* **86**, 21-34.

Porter, J. A., Young, K. E. and Beachy, P. A. (1996b). Cholesterol modification of hedgehog signaling proteins in animal development. *Science* **274**, 255-9.

Quirk, J., van den Heuvel, M., Henrique, D., Marigo, V., Jones, T. A., Tabin, C. and Ingham, P. W. (1997). The smoothened gene and hedgehog signal transduction in Drosophila and vertebrate development. *Cold Spring Harb Symp Quant Biol* **62**, 217-26.

Radhakrishnan, A., Sun, L. P., Kwon, H. J., Brown, M. S. and Goldstein, J. L. (2004). Direct binding of cholesterol to the purified membrane region of SCAP: mechanism for a sterol-sensing domain. *Mol Cell* **15**, 259-68.

Riddle, R. D., Johnson, R. L., Laufer, E. and Tabin, C. (1993). Sonic hedgehog mediates the polarizing activity of the ZPA. *Cell* **75**, 1401-16.

Rietveld, A., Neutz, S., Simons, K. and Eaton, S. (1999). Association of sterol- and glycosylphosphatidylinositol-linked proteins with Drosophila raft lipid microdomains. *J Biol Chem* **274**, 12049-54.

Rink, J., Ghigo, E., Kalaidzidis, Y. and Zerial, M. (2005). Rab conversion as a mechanism of progression from early to late endosomes. *Cell* **122**, 735-49.

Robbins, D. J., Nybakken, K. E., Kobayashi, R., Sisson, J. C., Bishop, J. M. and Therond, P. P. (1997). Hedgehog elicits signal transduction by means of a large complex containing the kinesin-related protein costal2. *Cell* **90**, 225-34.

Roelink, H., Augsburger, A., Heemskerk, J., Korzh, V., Norlin, S., Ruiz i Altaba, A., Tanabe, Y., Placzek, M., Edlund, T., Jessell, T. M. et al. (1994). Floor plate and motor neuron induction by vhh-1, a vertebrate homolog of hedgehog expressed by the notochord. *Cell* **76**, 761-75.

Rohatgi, R., Milenkovic, L., Corcoran, R. B. and Scott, M. P. (2009). Hedgehog signal transduction by Smoothened: pharmacologic evidence for a 2-step activation process. *Proc Natl Acad Sci U S A* **106**, 3196-201.

Rohatgi, R., Milenkovic, L. and Scott, M. P. (2007). Patched1 regulates hedgehog signaling at the primary cilium. *Science* **317**, 372-6.

Rudin, C. M., Hann, C. L., Laterra, J., Yauch, R. L., Callahan, C. A., Fu, L., Holcomb, T., Stinson, J., Gould, S. E., Coleman, B. et al. (2009). Treatment of medulloblastoma with hedgehog pathway inhibitor GDC-0449. *N Engl J Med* **361**, 1173-8.

Ruel, L., Rodriguez, R., Gallet, A., Lavenant-Staccini, L. and Therond, P. P. (2003). Stability and association of Smoothened, Costal2 and Fused with Cubitus interruptus are regulated by Hedgehog. *Nat Cell Biol* **5**, 907-13.

Ruiz i Altaba, A., Sanchez, P. and Dahmane, N. (2002). Gli and hedgehog in cancer: tumours, embryos and stem cells. *Nat Rev Cancer* **2**, 361-72.

Ryan, R. O. (1990a). Dynamics of insect lipophorin metabolism. *J Lipid Res* **31**, 1725-39.

Ryan, R. O., Wessler, A. N., Price, H. M., Ando, S. and Yokoyama, S. (1990b). Insect lipid transfer particle catalyzes bidirectional vectorial transfer of diacylglycerol from lipophorin to human low density lipoprotein. *J Biol Chem* **265**, 10551-5.

Schuck, S. and Simons, K. (2004). Polarized sorting in epithelial cells: raft clustering and the biogenesis of the apical membrane. *J Cell Sci* **117**, 5955-64.

Schwartz, G. G. and Skinner, H. G. (2007). Vitamin D status and cancer: new insights. *Curr Opin Clin Nutr Metab Care* **10**, 6-11.

Shapiro, J. P., Law, J. H. and Wells, M. A. (1988). Lipid transport in insects. *Annu Rev Entomol* **33**, 297-318.

Shelness, G. S. and Sellers, J. A. (2001). Very-low-density lipoprotein assembly and secretion. *Curr Opin Lipidol* **12**, 151-7.

Simons, K. and Gruenberg, J. (2000). Jamming the endosomal system: lipid rafts and lysosomal storage diseases. *Trends Cell Biol* **10**, 459-62.

Simons, K. and Ikonen, E. (1997). Functional rafts in cell membranes. *Nature* **387**, 569-72.

Sisson, J. C., Ho, K. S., Suyama, K. and Scott, M. P. (1997). Costal2, a novel kinesin-related protein in the Hedgehog signaling pathway. *Cell* **90**, 235-45.

Smelkinson, M. G., Zhou, Q. and Kalderon, D. (2007). Regulation of Ci-SCFSlimb binding, Ci proteolysis, and hedgehog pathway activity by Ci phosphorylation. *Dev Cell* **13**, 481-95.

Smolenaars, M. M., Kasperaitis, M. A., Richardson, P. E., Rodenburg, K. W. and Van der Horst, D. J. (2005). Biosynthesis and secretion of insect lipoprotein: involvement of furin in cleavage of the apoB homolog, apolipophorin-II/I. *J Lipid Res* **46**, 412-21.

Stanley, D. W. and Nelson, D. R. (1993). Insect lipids : chemistry, biochemistry, and biology.

Stegman, M. A., Goetz, J. A., Ascano, M., Jr., Ogden, S. K., Nybakken, K. E. and Robbins, D. J. (2004). The Kinesin-related protein Costal2 associates with membranes in a Hedgehog-sensitive, Smoothened-independent manner. *J Biol Chem* **279**, 7064-71.

Stenmark, H. (2009). Rab GTPases as coordinators of vesicle traffic. *Nat Rev Mol Cell Biol* **10**, 513-25.

Stone, D. M., Hynes, M., Armanini, M., Swanson, T. A., Gu, Q., Johnson, R. L., Scott, M. P., Pennica, D., Goddard, A., Phillips, H. et al. (1996). The tumour-suppressor gene patched encodes a candidate receptor for Sonic hedgehog. *Nature* **384**, 129-34.

Strigini, M. and Cohen, S. M. (1997). A Hedgehog activity gradient contributes to AP axial patterning of the Drosophila wing. *Development* **124**, 4697-705.

Strigini, M. and Cohen, S. M. (1999). Formation of morphogen gradients in the Drosophila wing. *Semin Cell Dev Biol* **10**, 335-44.

Strutt, H., Thomas, C., Nakano, Y., Stark, D., Neave, B., Taylor, A. M. and Ingham, P. W. (2001). Mutations in the sterol-sensing domain of Patched suggest a role for vesicular trafficking in Smoothened regulation. *Curr Biol* **11**, 608-13.

Sundermeyer, K., Hendricks, J. K., Prasad, S. V. and Wells, M. A. (1996). The precursor protein of the structural apolipoproteins of lipophorin: cDNA and deduced amino acid sequence. *Insect Biochem Mol Biol* **26**, 735-8.

Tabata, T. (2001). Genetics of morphogen gradients. *Nat Rev Genet* **2**, 620-30.

Tabata, T. and Kornberg, T. B. (1994). Hedgehog is a signaling protein with a key role in patterning Drosophila imaginal discs. *Cell* **76**, 89-102.

Taipale, J., Chen, J. K., Cooper, M. K., Wang, B., Mann, R. K., Milenkovic, L., Scott, M. P. and Beachy, P. A. (2000). Effects of oncogenic mutations in Smoothened and Patched can be reversed by cyclopamine. *Nature* **406**, 1005-9.

Taipale, J., Cooper, M. K., Maiti, T. and Beachy, P. A. (2002). Patched acts catalytically to suppress the activity of Smoothened. *Nature* **418**, 892-7.

Teleman, A. A., Strigini, M. and Cohen, S. M. (2001). Shaping morphogen gradients. *Cell* **105**, 559-62.

Torroja, C., Gorfinkiel, N. and Guerrero, I. (2004). Patched controls the Hedgehog gradient by endocytosis in a dynamin-dependent manner, but this internalization does not play a major role in signal transduction. *Development* **131**, 2395-408.

Tseng, T. T., Gratwick, K. S., Kollman, J., Park, D., Nies, D. H., Goffeau, A. and Saier, M. H., Jr. (1999). The RND permease superfamily: an ancient, ubiquitous and diverse family that includes human disease and development proteins. *J Mol Microbiol Biotechnol* **1**, 107-25.

Tufail, M. and Takeda, M. (2009). Insect vitellogenin/lipophorin receptors: molecular structures, role in oogenesis, and regulatory mechanisms. *J Insect Physiol* **55**, 87-103.

van den Brink, G. R. (2007). Hedgehog signaling in development and homeostasis of the gastrointestinal tract. *Physiol Rev* **87**, 1343-75.

van den Heuvel, M. and Ingham, P. W. (1996a). smoothened encodes a receptor-like serpentine protein required for hedgehog signalling. *Nature* **382**, 547-51.

Van der Horst, D. J. (1990). Lipid transport function of lipoproteins in flying insects. *Biochim Biophys Acta* **1047**, 195-211.

van Meer, G., Voelker, D. R. and Feigenson, G. W. (2008). Membrane lipids: where they are and how they behave. *Nat Rev Mol Cell Biol* **9**, 112-24.

Vance, J. E. and Vance, D. (2002). Biochemistry of lipids, lipoproteins and membranes. 4th edition.

Von Ohlen, T. and Hooper, J. E. (1997). Hedgehog signaling regulates transcription through Gli/Ci binding sites in the wingless enhancer. *Mech Dev* **68**, 149-56.

Wang, Q. T. and Holmgren, R. A. (1999). The subcellular localization and activity of Drosophila cubitus interruptus are regulated at multiple levels. *Development* **126**, 5097-106.

Wang, Y., Zhou, Z., Walsh, C. T. and McMahon, A. P. (2009). Selective translocation of intracellular Smoothened to the primary cilium in response to Hedgehog pathway modulation. *Proc Natl Acad Sci U S A* **106**, 2623-8.

Wasserman, R. H., Henion, J. D., Haussler, M. R. and McCain, T. A. (1976). Calcinogenic factor in Solanum malacoxylon: evidence that it is 1,25-dihydroxyvitamin D3-glycoside. *Science* **194**, 853-5.

Watari, H., Blanchette-Mackie, E. J., Dwyer, N. K., Watari, M., Neufeld, E. B., Patel, S., Pentchev, P. G. and Strauss, J. F., 3rd. (1999). Mutations in the leucine zipper motif and sterol-sensing domain inactivate the Niemann-Pick C1 glycoprotein. *J Biol Chem* **274**, 21861-6.

Wendler, F., Franch-Marro, X. and Vincent, J. P. (2006). How does cholesterol affect the way Hedgehog works? *Development* **133**, 3055-61.

Wiegandt, H. (1992). Insect glycolipids. *Biochim Biophys Acta* **1123**, 117-26.

Williams, E. H., Pappano, W. N., Saunders, A. M., Kim, M. S., Leahy, D. J. and Beachy, P. A. (2008). Dally-like core protein and its mammalian homologues mediate stimulatory and inhibitory effects on Hedgehog signal response. *Proc Natl Acad Sci U S A* **107**, 5869-74.

Wojtanik, K. M. and Liscum, L. (2003). The transport of low density lipoprotein-derived cholesterol to the plasma membrane is defective in NPC1 cells. *J Biol Chem* **278**, 14850-6.

Xie, Z. and Bikle, D. D. (1998). Differential regulation of vitamin D responsive elements in normal and transformed keratinocytes. *J Invest Dermatol* **110**, 730-3.

Yauch, R. L., Gould, S. E., Scales, S. J., Tang, T., Tian, H., Ahn, C. P., Marshall, D., Fu, L., Januario, T., Kallop, D. et al. (2008). A paracrine requirement for hedgehog signalling in cancer. *Nature* **455**, 406-10.

Zhang, C., Williams, E. H., Guo, Y., Lum, L. and Beachy, P. A. (2004). Extensive phosphorylation of Smoothened in Hedgehog pathway activation. *Proc Natl Acad Sci U S A* **101**, 17900-7.

Zhang, Q., Zhang, L., Wang, B., Ou, C. Y., Chien, C. T. and Jiang, J. (2006). A hedgehog-induced BTB protein modulates hedgehog signaling by degrading Ci/Gli transcription factor. *Dev Cell* **10**, 719-29.

Zhang, W., Zhao, Y., Tong, C., Wang, G., Wang, B., Jia, J. and Jiang, J. (2005). Hedgehog-regulated Costal2-kinase complexes control phosphorylation and proteolytic processing of Cubitus interruptus. *Dev Cell* **8**, 267-78.

Zhao, Y., Tong, C. and Jiang, J. (2007). Hedgehog regulates smoothened activity by inducing a conformational switch. *Nature* **450**, 252-8.

Zhou, J. X., Jia, L. W., Liu, W. M., Miao, C. L., Liu, S., Cao, Y. J. and Duan, E. K. (2006). Role of sonic hedgehog in maintaining a pool of proliferating stem cells in the human fetal epidermis. *Hum Reprod* **21**, 1698-704.

Zhu, A. J., Zheng, L., Suyama, K. and Scott, M. P. (2003). Altered localization of Drosophila Smoothened protein activates Hedgehog signal transduction. *Genes Dev* **17**, 1240-52.

10   ACKNOWLEDGEMENT

I am endlessly grateful to Suzanne Eaton, who not only gave me a possibility to perform this work in her lab, but who was also incredibly supportive and motivating – both professionally and personally – through and through. Thank you, Suzanne, for teaching me to believe "at least six impossible things".

I gratefully acknowledge the Max Planck Institute for Molecular Cell Biology and Genetics for giving me a unique opportunity to work in such a special place, where true science is done happily.

Many great thanks to Eli Knust, Kai Simons and Marino Zerial for fruitful discussions, useful comments and support during the publication of my scientific manuscript.

My special thanks to Kai Simons and Temo Kurzchalia – my TAC members, who always provided me with highly motivating thoughts. Thank you – for your crucial advice throughout.

I am greatly thankful to all the members of the Eaton lab – the past and the present. Thank you for countless coffees and late hour smiles – you made these years brighter and cheery for me.

Thank you, Daniela Panáková, for handing over the project – you taught me a lot about science and science politics.

My very special thanks to you, Ali. Thank you for your endless support, for always being there, for everything... You have been amazing to me.

I am eternally grateful to my whole big family, who – although far away – always were at my side and strongly believed in me. Thank you for telling me at the age of six, that I would make a great scientist some day.

I am infinitely thankful to all my friends, close and further away, for all the amazing words said, all the inspiration given and all the thousands of hugs and smileys. Everyone of you has a special place in my heart.

Thank you, папа Гият и мама Иида, for being the world to me.

Christian, thank you for your overwhelming care and belief in me, no matter what, "for better and for worse", thank you for making me 'your everything'. If it were not for you, this work would not be possible.

11 PUBLICATION

This work has been published as

Khaliullina, H., Panakova, D., Eugster, C., Riedel, F., Carvalho, M. and Eaton, S. (2009). Patched regulates Smoothened trafficking using lipoprotein-derived lipids. *Development* **136**, 4111-21.

Die VDM Verlagsservicegesellschaft sucht für wissenschaftliche Verlage abgeschlossene und herausragende

Dissertationen, Habilitationen, Diplomarbeiten, Master Theses, Magisterarbeiten usw.

für die kostenlose Publikation als Fachbuch.

Sie verfügen über eine Arbeit, die hohen inhaltlichen und formalen Ansprüchen genügt, und haben Interesse an einer honorarvergüteten Publikation?

Dann senden Sie bitte erste Informationen über sich und Ihre Arbeit per Email an *info@vdm-vsg.de*.

Sie erhalten kurzfristig unser Feedback!

VDM Verlagsservicegesellschaft mbH
Dudweiler Landstr. 99
D - 66123 Saarbrücken

Telefon +49 681 3720 174
Fax +49 681 3720 1749

www.vdm-vsg.de

Die VDM Verlagsservicegesellschaft mbH vertritt

Printed by Books on Demand GmbH, Norderstedt / Germany